Autonomous Vehicle Navigation

From Behavioral to Hybrid
Multi-Controller Architectures

Autonomous Vehicle Navigation

From Behavioral to Hybrid Multi-Controller Architectures

Lounis Adouane

Institut Pascal - Polytech
Clermont-Ferrand, France

CRC Press
Taylor & Francis Group
Boca Raton London New York

CRC Press is an imprint of the
Taylor & Francis Group, an **informa** business

A CHAPMAN & HALL BOOK

CRC Press
Taylor & Francis Group
6000 Broken Sound Parkway NW, Suite 300
Boca Raton, FL 33487-2742

Visit the Taylor & Francis Web site at
http://www.taylorandfrancis.com

and the CRC Press Web site at
http://www.crcpress.com

To my parents (Hayat and Larbi)
To my family
To my wife and our two angels (Tanirt and Yani)

Contents

List of Figures

List of Tables

Foreword

This manuscript will emphasize the investigated scientific research themes addressed throughout our developments since my incorporation in September 2006 at Polytech Clermont-Ferrand – Institut Pascal UMR CNRS 6602 (France) as an associate professor. Obviously, it will not detail exhaustively all the developments undertaken since 2006, but will only focus on the most important achievements/results while highlighting the innovative scientific methodology leading to the different outcomes.[1] The presented researches are focused on the way to **increase the autonomy** of mobile **mono robot** as well as **Multi-Robot Systems** (MRS) to **achieve complex tasks**. More precisely, the main objective is to emphasize the developed **generic control architectures** in order to enhance the **safety**, **flexibility** and the **reliability** of **autonomous navigations in complex environments** (e.g., cluttered, uncertain and/or dynamic). The proposed control architectures (**decision/action**) have been addressed through three closely related elements: **task modeling**; **planning** and finally the **control** aspect. Among the main ideas developed in this manuscript are those related to the potentiality of using **multi-controller architectures**.[2] Indeed, using this kind of control permits us to break the complexity of the overall tasks to be carried out and therefore allows a **bottom-up** development. This will imply the development of appropriate reliable elementary controllers (**obstacle avoidance, target reaching/tracking, formation maintaining,** etc.), but also the proposition of appropriate mechanisms to manage the interaction of these multi-controller architectures while ensuring the respect of different constraints and enhancing metrics/criteria linked to the safety, flexibility and reliability of the overall control.

Although the developed concepts/methods/architectures could be applied for different domains (such as service robotics or agriculture), the **transportation domain** remains the privileged target. Applications include the transportation of persons (private car or public transport) as well as merchandise transportation (in warehouses or ports for instance). The different proposals will be applied for simple robotic entities (like Khepera® robots modeled as unicycles) as much as for larger ones (like VIPALAB® vehicles modeled as tricycles). The theoretical aspect will take a part of the manuscript, but several simulations and experiments will be given to demonstrate the efficiency of the adopted approaches.

---

[1] The details could be shown in the referenced papers, supervised PhD thesis, project manuscripts, etc.

[2] Well-known initially in the literature as behavioral control architectures (cf. section 1.4, page 15).

Acknowledgments

This book would not have been possible without the help and support of my family members, many friends, students, and colleagues in the field. It would be impossible to list all of them here.

First of all, it is a great pleasure to acknowledge the people who have closely collaborated with me in the research contained in this book. A special acknowledgment goes to all the current and past PhD students who worked intensively with me during these last years; in alphabetical order, let me cite: Ahmed Benzerrouk, Suhyeon Gim, Bassem Hichri, Guillaume Lozenguez, Mehdi Mouad and José Miguel Vilca Ventura.

I would also like to thank all my colleagues at Institut Pascal/IMobS3 for their valuable exchange and support, among them, while being certainly not exhaustive: Omar Ait-Aider, Nicolas Andreff, François Berry, Pierre Bonton, Roland Chapuis, Thierry Chateau, Jean-Pierre Derutin, Michel Dhome, Jean-Christophe Fauroux, Philippe Martinet, Youcef Mezouar and Benoit Thuilot.

My thanks also go to my colleagues abroad, mainly to Djamel Khadraoui (LIST - Luxembourg), Sukhan Lee (ISRI - South Korea), Antonios Tsourdos and Andrzej Ordys (respectively Cranfield and Kingston universities - United Kingdom), Carlos SAGUES (I3A - Spain), Helder Araujo (CRVLAB - Portugal) and also to colleagues who have kindly accepted to give me the authorization to use some materials (images, references, etc.) of their valuable works, among them: Prof. Marco Dorigo - Université libre de Bruxelles (Belgium); Prof. Ernst D. Dickmanns - Bundeswehr University of Munich (Germany); Prof. Michael R. M. Jenkin - York University (Canada); Prof. Satoshi Murata and Prof. Yasuhisa Hirata from Tohoku University (Japan).

Special thanks go to several people for their contributions to the development and production of the book at CRC Press/Taylor and Francis, including Sarah Chow (Associate Acquisitions Editor), Randi Cohen (Senior Acquisitions Editor), Donley Amber (Project Coordinator) and Michael Davidson (Project Editor).

It is important to mention that a great deal of the research represented within this manuscript was funded by French National Research Agency (ANR) and ADEME (Agence de l'Environnement et de la Maîtrise de l'Energie) through Investissements d'Avenir program of the French government.

Author Biography

Lounis Adouane is an associate professor since 2006 at the Institut Pascal–Polytech Clermont-Ferrand in France. He received an MS in 2001 from IRCCyN–ECN Nantes, where he worked on the control of legged mobile robotics. In 2005, he obtained a PhD in automatic control from FEMTO-ST laboratory–UFC Besançon. During his PhD studies he deeply investigated the field of multi-robot systems, especially those related to bottom-up and reactive control architectures. After that, he joined in 2005 Ampère laboratory–INSA Lyon and studied hybrid (continuous/discrete) control architectures applied to cooperative mobile robot arms. In 2014, he spent 6 months as a visiting professor in two robotics laboratories at Cranfield and Kingston universities (United Kingdom). Dr. Adouane's main research and teaching activities are linked to robotics, automatic control and computer science. His current research topics are related to both *autonomous navigation of mobile robots in complex environments* and *cooperative control architectures for multi-robot systems*. Since 2006, he has authored/coauthored more than 60 international references dealing mainly with the following keywords: autonomous mobile robots/vehicles; control of complex systems; multi-controller architectures; hybrid (continuous/discrete) and hybrid (reactive/cognitive) control architectures; Lyapunov-based synthesis and stability; obstacle avoidance (static and dynamic); limit-cycle approach; target reaching/tracking; cooperative multi-robot systems; navigation in formation (virtual structure, leader-follower); cooperative exploration task; cooperative transportation task; task allocation; auction coordination; kinematic constraints; constrained control; optimal planning; continuous curvature path; clothoids composition; velocity planning; waypoints generation; multi-criteria optimization; artificial intelligence (such as Markov decision process, multi-agent system, fuzzy logic, etc.); multi-robot/agent simulation.

General Introduction

GENESIS OF THE RESEARCH WORKS

During my PhD thesis [Adouane, 2005], achieved in the micro-robotics team at LAB (Labortaoire d'Automatique de Besançon, France), the research objective was to control a group of minimalist mobile robots, called ALICE [Caprari, 2003] (with a dimension of 2cm $\times$ 2cm $\times$ 2cm) to perform, among others, the CBPT[3] (Cooperative Box-Pushing Task, cf. Figure 1(a)). The constraints imposed by the use of these minimalist structures as well as the nature of the achieved cooperative task, which aims to control the navigation and the interaction of a swarm of mobile mini-robots, led us to develop several mechanisms/ideas to deal with this highly dynamic system. Indeed, the interaction of a swarm of mini-robots in the immediate vicinity of the box to push is very high and needs to be addressed without neither high cognition/planning (cf. section 1.3.2, page 11) aspects nor centralized control (cf. section 1.3.3, page 13) [Adouane, 2005]. Therefore, fully reactive and decentralized behavioral control architectures have been proposed to take into account the different constraints linked to the control of this highly dynamic swarm of robots. More precisely, a Hierarchical Action Selection Process (HASP) was proposed which allows us to coordinate with stimuli-response mechanism, the activity of the elementary behaviors/controllers composing the proposed architectures. The HASP has been, thereafter, improved by integrating mechanisms of fusion of actions and a mechanism of dynamical gains adaptation [Adouane and Le Fort-Piat, 2005], to obtain the Hybrid-HASP [Adouane and Le-Fort-Piat, 2004]. This last process of coordination is more flexible, intuitive and scalable than the basic HASP, and it has been proved to be strongly adapted to control highly dynamic multi-robot systems. This process allows us, at the level of the robot, to coordinate in a hierarchical and flexible manner the activity of a set of elementary controllers (behaviors), and at the level of the

---

[3]In the field of swarm robotics, the CBPT is among the privileged complex task, in order to study the relevance of reactive and decentralized control architectures [Parker, 1999], [Yamada and Saito, 2001], [Ahmadabadi and Nakano, 2001], [Baldassarre et al., 2003], [Muñoz, 2003].

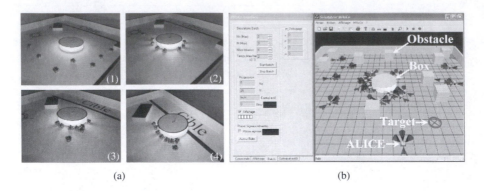

(a)   (b)

FIGURE 1   (See color insert) (a) Experimentation of the proposed control architecture with 8 mini-robots ALICE pushing a cylindrical object to a final
assigned area; (b) *MiRoCo* simulator.

group of robots, the coordination of the robot's interactions for reaching global objectives and desired mass effects. Otherwise, specific low-level communication, called
altruistic behaviors, reproducing the simple interaction (attraction/repulsion) of indi-
viduals constituting societies of insects [Bonabeau et al., 1999] were integrated into
the proposed control architectures in order to improve the efficiency of robots' coordination [Adouane, 2005].

The validation of the proposed mechanisms of control was made through actual experiments (cf. Figure 1(a)) but more intensively according to statistical
studies done on a large number of data obtained thanks to MiRoCo[4] simulator (cf. Figure 1(b)). The performed statistical studies show, among others, the
existence of an optimal number of robots to achieve the CBPT and underline
the importance of implicit communications induced by the altruistic behaviors
[Adouane and Le-Fort-Piat, 2004]. It is important to emphasize that with this kind
of approach, followed during the PhD thesis [Adouane, 2005], there is no other analytic technique to prove the actual reliability of the proposed control/strategy. The
step consisting of using statistical study in order to prove the efficiency of the proposals is in general mandatory in these kinds of approaches [Bonabeau et al., 1999].

Although if the obtained results during the PhD are efficient to control a highly
dynamic multi-robot system, the lack of accurate analytic analysis reduces drastically
the scope of possible use of the already proposed control architectures, especially if
the targeted tasks imply close interaction between the robots and humans (e.g., transportation or service robotics tasks) or industrial applications (e.g., automatic warehouse management/manipulation/transportation by a group of mobile robots). It has
been decided thereafter to improve the features of these multi-controller architec-

---

[4]*MiRoCo* (for Mini Robotique Collective, cf. ref [Adouane, 2005, chapter 7]) is a reliable and 3D
simulator dedicated in general to cooperative mobile robotics. *MiRoCo* gives a very good approximation
of the different physical constraints linked to the interaction of the robots between them and with their
environment.

tures, which still have large potentialities (cf. section 1.4, page 15), while permitting analytic and accurate stability/reliability analysis. This could be reached while introducing more automatic control theory and while better mastering the elementary developed controllers and their interactions, in order to actually attest to the reliability of the overall control architecture.

IMPORTANT INVESTIGATED DOMAINS AND MANUSCRIPT STRUCTURE

The main ideas developed in this book are related to the potentialities of using multi-controller architectures[5] to tend ineluctably toward fully autonomous robot navigation even in highly dynamic and cluttered environments. Indeed, using this kind of control permits us to break up the complexity of the overall tasks to be carried out and therefore allows a bottom-up development. It will be shown in this manuscript how the proposed techniques, concepts and methodologies can address different complex mobile robot tasks. This will imply the development of appropriate reliable elementary controllers (obstacle avoidance, target reaching, formation maintaining, etc.), but also the proposition of appropriate mechanisms to manage the interaction of these multi-controller architectures while ensuring the respect of different constraints and enhancing metrics/criteria linked to the safety, flexibility and reliability of the overall proposed control architectures.

Furthermore, in order to enhance the autonomy of mobile robots, several investigated works will be presented in this book, dealing with: modeling of sub-tasks; reliable obstacle avoidance; appropriate stable control laws for target reaching/tracking; short-term and long-term trajectories/waypoints planning; navigation through sequential waypoints; cooperative control and interaction of a group of mobile robots. More precisely, this manuscript is organized into 6 chapters:

- **Chapter 1** introduces briefly the domain of autonomous mobile robotics while highlighting its main achievements/challenges. It will also emphasize also the main concepts/paradigms/motivations/definitions used throughout the text. The objective is to clarify them in order to simplify the different explanations and developments used in the rest of the book. Among them let us cite: the boundary limit between planning and control; interest and most challenging aspects linked to multi-controller architectures; the notion of reactive/cognitive, centralized/decentralized,[6] flexibility, stability and reliability of the developed control architectures.

- **Chapter 2** is devoted to an important navigation function corresponding to obstacle avoidance controller. Thereby, a safe and flexible obstacle avoidance controller, based on Limit-Cycles, for autonomous navigation in cluttered environments will be presented. This chapter will also introduce important ele-

---

[5]Well-known initially in the literature as behavioral control architectures (cf. section 1.4, page 15).

[6]It is to be noted that in literature, cognitive and decentralized are, respectively, also called deliberative and distributed control architectures.

mentary building blocks characterizing the different multi-controller architectures developed throughout this manuscript. A brief description of the methodology to detect and to characterize obstacles in the environment will also be presented.

- **Chapter 3** focuses on the proposed Hybrid$_{CD}$ (Continuous/Discrete) multicontroller architectures for online mobile robot navigation in cluttered environments. The developed stable control laws for target reaching/tracking will be presented. An important part of this chapter emphasizes how to obtain stable and smooth switching between the different elementary controllers composing the proposed architectures.

- **Chapter 4** focuses on the proposed Hybrid$_{RC}$ (Reactive/Cognitive) control architecture permitting us to simply manage the activation of reactive and cognitive navigation according to the environment context (uncertain or not, dynamic or not, etc.). This architecture is based among others on the use of the homogeneous set-points definition coupled with appropriate control law shared by all the controllers. This chapter will pay attention to the proposed planning methods, mainly the one based on PELC for car-like robots.

- **Chapter 5** emphasizes the fact that it is not absolutely mandatory (as commonly admitted and broadly used in the literature) to have a predetermined trajectory to be followed by a robot to perform reliable and safe navigation in an urban and/or cluttered environment. In this chapter, a new definition of the navigation task, using only discrete waypoints in the environment, will be presented and applied for an urban electric vehicle. This approach permits us to reduce the computational costs and leads to even more flexible navigation with respect to traditional approaches (mainly if the environment is cluttered and/or dynamic).

- **Chapter 6** is dedicated to the control of multi-robot systems. The focus will be on dynamic multi-robot navigation in formation and on the cooperative strategies to perform safe, reliable and flexible navigation. An overview of other addressed multi-robot tasks (such as "co-manipulation and transportation" and "exploration under uncertainty") will also be briefly presented.

A general conclusion and prospects are given at the end of the book.

CHAPTER 1

Global concepts/challenges related to the control of intelligent mobile robots

CONTENTS

T HIS CHAPTER provides an overview of the autonomous mobile robotics domain while highlighting its main achievements/challenges. It will also emphasize the main concepts/paradigms/motivations/definitions used throughout the rest of the manuscript. The objective is to clarify them in order to simplify the different explanations and developments given in the coming chapters.

1.1 AUTONOMOUS/INTELLIGENT MOBILE ROBOTS

During the last decades, research investigations, related to the field of autonomous navigation of mobile robots become more and more important. It would be illusory to try to cover all the different research / developments / projects related to autonomous mobile robots. In fact, several laboratories / companies / industries / start-ups throughout the world are involved in this useful area.

Robotic mobilities Several kinds of robot mobilities exist and they can be classified according to their work space/environment, among the most representative, let us cite:

- Underwater robots which are commonly called Autonomous Underwater Vehicles (AUVs) (cf. Figure 1.1(a)).

- Boat robots, called Unmanned Surface Vehicles (USVs) (cf. Figure 1.1(b)).

- Aerial robots which are commonly called Unmanned Aerial Vehicles (UAVs) or drones (cf. Figure 1.1(c)).

- Ground robots which are commonly called Unmanned Ground Vehicles (UGVs). Furthermore, the UGV can use different devices to move such as legs (cf. Figure 1.1(d)), wheels (cf. Figure 1.1(e)) or specialized wheels [Siegwart et al., 2011] [Seeni et al., 2010] such as tracks wheels [Wu et al., 2014] or the one used by the robot Curiosity (cf. Figure 1.1(f)).

The different works developed in this manuscript are dedicated to UGVs with wheels. The term mobile robot will always designate, in what follows, a UGV and if not, the mobile robot structure will be explicitly defined in the text. More specifically, the focus will be on the control of UGVs (mono or multi-robot entities) to perform **autonomous navigation** even in **highly cluttered and dynamic environments** (cf. section 2.1, page 26). In the adopted methodology, it is taken as principle that if the proposed control architecture can be reliable in that kind of complex and constrained environment, it could be obviously even more reliable in a less constrained one.

Autonomous navigation of UGVs is used in different tasks/domains, for instance: area surveillance [Borja et al., 2013] [Memon and Bilal, 2015], mapping of unknown environments [Lategahn et al., 2011] [Fernández-Madrigal and Claraco, 2013], human search and rescue [Murphy, 2012] [Simpkins and Simpkins, 2014], space exploration [Seeni et al., 2010] [Alfraheed and Al-Zaghameem, 2013], military [Voth, 2004] [Springer, 2013], agriculture [Guillet et al., 2014] [Lu and Shladover, 2014], service robotics [Goodrich and Schultz, 2007] [Güttler et al., 2014] or transportation [Takahashi et al., 2010] [Vilca et al., 2015a].

Although the developed concepts/methods/architectures could be applied for the different tasks/domains given above, the transportation domain remains our main privileged target. This transportation can touch people (private car or public transport) as well as goods transportation (in warehouses or ports, for instance).

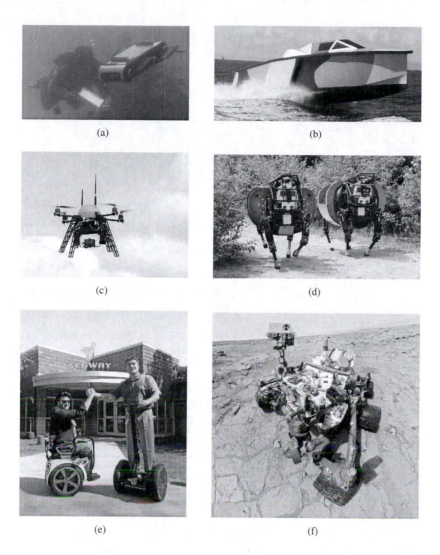

FIGURE 1.1 (See color insert) Different kinds of mobile robot locomotion. (a) AQUA® bio-inspired AUV (KROY version) [Speers and Jenkin, 2013], (b) Piranha® USV from Zyvex® company, (c) UAV from Fly-n-Sense® company, (d) two LS3s from Boston Dynamics®, (e) the first two-wheel self-balancing chair Genny®, inspired by Segway®, (f) Robot Curiosity (from NASA) exploring Mars planet.

Short historical aspects The vision to have driverless vehicles is not new. Indeed, since 1939, in the Futurama Exhibition held in New York and sponsored by General Motors, the idea to have radio-controlled cars in the motorway was announced.

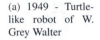

(a) 1949 - Turtle-like robot of W. Grey Walter   (b) 1967 - Shakey mobile robot   (c) 1987 - VaMoRs Autonomous Van

FIGURE 1.2   (See color insert) UGV's short historic aspect.

Before we explore more this important domain, let us give some important historic steps related to the first and determining UGV structures/abilities:

- 1949: The first mobile robots were built by W. Grey Walter in 1949. They have three wheels, and are turtle-like robots (Elmer and Elsie turtles) with light and touch sensors, drive and steering motors, and two vacuum tube analog computers [Walter, 1953] (cf. Figure 1.2(a)). This system allowed the turtles to wander in a room and return to a recharging station.

- 1967: Stanford Research Institute (SRI) develops Shakey mobile robot [Wilber, 1972] (cf. Figure 1.2(b)), the first sophisticated mobile robot which has its own embedded perceptive and control aspects. The first mobile robot with the same sophistication has been developed in France at LAAS (Laboratoire d'Analyse et d'Architecture des Systèmes) in 1977. This robot is called Hilare.

- 1987: University Bundeswehr Munich (UniBwM) experimented with the first driverless car, called VaMoRs (**V**ersuchsfahrzeug für **a**utonome **Mo**bilität und **R**echnersehen, cf. Figure 1.2(c)) in motorway without traffic (velocity of ± 60 km/h). VaMoRs with many special features, has served as test platform of UniBw Munich from 1985 to 2004 with three generations of vision systems and has demonstrated many, firsts' in the field of road vehicles capable of visual perception. This autonomous navigation was done using notably vision and probabilistic methods. Autonomous mobile robotics reach a milestone and automotive industries increase their interest for this new applications.

Short overview of interesting former/current projects   Without being exhaustive to cite the multitude of projects raised since the 80's around UGVs, it is important to notice that several relevant projects from Europe and the USA have been developed around driverless cars. Among the most important, which had a great impact on the scientific community as well as industries/companies, let us cite: from 1987 to 1995, the Eureka PROMETHEUS project (PROgraMme for a European Traffic of Highest Efficiency and Unprecedented Safety) [Williams, 1988] [EUREKA, 1995] which was a very large R&D project for driverless cars and involved numerous universities and car manufacturers. In 2004 the first DARPA (Defense Advanced Research Projects Agency) Grand Challenge was launched: organized with no traffic, dirt roads, driven by a predefined route and with obstacles known in advance [DARPA, 2015]; 3 years after (in 2007) the Urban Grand Challenge race [Thrun et al., 2007] [Buehler et al., 2009] had great success. The race involved a 96 km urban area. Rules included obeying all traffic regulations while negotiating with other traffic and obstacles and merging into traffic. The important rate of success of this challenge opens a large spectrum to define the actual possibilities of the driverless car for mid- and long-term prospects.

In recent years, the development of a fully autonomous vehicle in the transportation field has received even more attention from different countries [Burns, 2013]. One of the most important events which has drawn a lot of public attention with full autonomy is in 2010, the Google driverless car [Thrun, 2011] (cf. Figure 1.3(a)) or in 2013 the BRAiVE[1] car from Parma University (Italy). These two vehicles permit fully autonomous driving in different contexts (rural / free-way / urban).

Several other interesting projects involve a fleet of vehicles in order for instance: to **reduce energy consumption** while enhancing the fleet aerodynamics (such as cyclists) and the possibility to avoid much accelerating and braking (since all of the fleet have the same speed); this cooperative and automated driving system may also provide promising solutions to traffic congestion (reduce/condensate the occupancy space) or safety issues [GCDC, 2016]. Among these interesting projects let us cite: SARTRE[2] project (cf. Figure 1.3(b)) which involved seven European partners and was finalized in 2012; in 2016 we will see the second edition of Grand Cooperative Driving Challenge (GCDC [GCDC, 2016], cf. Figure 1.3(c)). The first GCDC was held in May 2011 in Helmond, the Netherlands. The GCDC 2011 was mainly focused on the ability to perform longitudinal control of the vehicles (platooning). In the 2016 edition, in addition to lateral control (steering), cooperative driving will be the main focus. Challenges awaiting the participants include the ability to merge platoons and to join a busy road on a T intersection without driver intervention. Nowadays several important automotive manufacturers like Mercedes, GM, Ford, Daimler, Audi, BMW or Nissan,[3] etc. announce to sell a driverless car at the mid-term horizon (less than 10 years), but before that, important challenges must be resolved (cf. section 1.2).

It is important to notice that the driverless car is not only synonym of a car as we commonly know but with the automation of its displacement functions. In fact,

---

[1]http://www.braive.vislab.it/, consulted January 2015.

[2]SARTRE (SAfe Road TRains for the Environment) http://www.sartre-project.eu/, consulted January 2015.

[3]"The realization of the Autonomous Drive system is one of our greatest goals, because Zero Fatalities stands alongside Zero Emissions as major objective of Nissan's R&D," says Mitsuuhiko Yamasahita (VP R&D NISSAN).

(a) Google® Car

(b) SARTRE project

(c) GCDC project

(d) VIPALAB® vehicles

FIGURE 1.3 (See color insert) Different autonomous UGVs projects in several environments.

in parallel with the developments of this area by automotive industries and certain laboratories, another generation of UGVs like VIPALAB (cf. Figure 1.3(d)) aims also to autonomously transport passengers but in a more restricted area like midtown or inside big companies, amusement parks, airports, etc. which need autonomous shuttles between their different area. This specific autonomous navigation function is among the most important applications started at Institut Pascal since 2000. Although if the environment of navigation is generally delimited and the dynamic of UGV evolution is not the same as for the Google car for instance, nevertheless an important part of the autonomous navigation issues are shared. Indeed, this kind of UGV must, like the Google car, navigate autonomously while taking into account the different events (e.g., traffic light, obstructing objects, etc.). Further, a lot of work has been done using these UGVs for autonomous navigation in formation with different shapes (platooning among them), as is the case of the SafePlatoon[4] project where the targeted applications are linked to urban, military and agriculture environments. VIPALAB vehicles constitute one of our privileged platforms to experiment with our developments (cf. chapter 5 and chapter 6).

---

[4]http://web.utbm.fr/safeplatoon/, consulted January 2015.

1.2 OVERVIEW OF THE CHALLENGES RELATED TO FULLY AUTONOMOUS NAVIGATION

The objective of autonomous vehicles is to improve the quality of life, with pollution reduction (greenhouse gas) and accident prevention [Litman, 2013] [Fagnant and Kockelman, 2015] [Vine et al., 2015] (for instance, 93% of road traffic accidents are caused by human error [Yeomans, 2010]). Furthermore, knowing the vision of futuristic smart cities with, for instance, shared green and autonomous cars (on-demand offers[5]) in midtown (to reduce among other traffic congestion), the area of automotive is currently in full mutation and interests therefore industrialists as well as scientists.

For a formal definition of cars autonomy, reference [SAE, 2015] gives an interesting classification using 6 levels. Figure 1.4 shows the 6 defined levels, from the lowest (level 0 "No Automation") to the highest (level 5 "Full Automation" door-to-door feature). The work developed in this manuscript targets levels 4 and 5.

To obtain a fully autonomous mobile robot it is important to master 3 complementary phases: the perception/localization, the decision and the action. The perception/localization phase [Royer et al., 2007] must built an even simpler model of the robot's environment and must permit either local or global robot localization according, for instance, to local obstacles or global environment; the decision uses this model/localization to generate the appropriate set-points to achieve the assigned task and finally the action turns these set-points into corresponding commands (using appropriate control laws) for the robot's actuators [Adouane, 2005]. Our works focus on the decision and action phases.

In order to enhance the vehicle's autonomy, the communication aspect (cf. Figure

---

[5]It is to be noted that individual cars are used not much more than 5% of their lifetime.

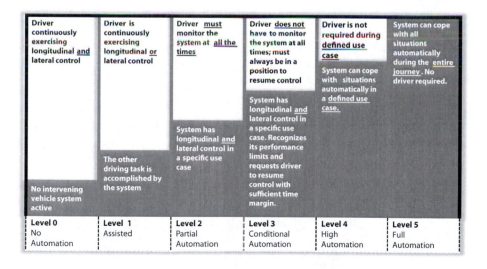

Level 0	Level 1	Level 2	Level 3	Level 4	Level 5
Driver continuously exercising longitudinal and lateral control	Driver is continuously exercising longitudinal or lateral control	Driver must monitor the system at all the times	Driver does not have to monitor the system at all times; must always be in a position to resume control	Driver is not required during defined use case	System can cope with all situations automatically during the entire journey. No driver required.
	The other driving task is accomplished by the system	System has longitudinal and lateral control in a specific use case	System has longitudinal and lateral control in a specific use case. Recognizes its performance limits and requests driver to resume control with sufficient time margin.	System can cope with situations automatically in a defined use case.	
No intervening vehicle system active					
Level 0 No Automation	Level 1 Assisted	Level 2 Partial Automation	Level 3 Conditional Automation	Level 4 High Automation	Level 5 Full Automation

FIGURE 1.4   Different levels of a car's autonomy [SAE, 2015].

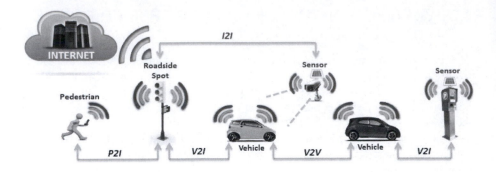

FIGURE 1.5 Different communications features to enhance a car's autonomy/safety. Vehicular Ad-hoc NETworks (VANET): Vehicle-to-Vehicle (V2V), Vehicle-to-Infrastructure (V2I), Infrastructure-to-Infrastructure (I2I) and Pedestrian-to-Infrastructure (P2I) [VANET, 2015] [Papadimitratos et al., 2008] [Toutouh et al., 2012].

1.5) must also be mastered and secured between the UGVs and their environments (cf. Figure 1.5, e.g., traffic light, pedestrian, intersection, etc.). In fact, exchange of information enables cooperation between UGVs, and between UGVs and roadside systems (the infrastructure). Through access to, for example, early warnings on up-coming traffic situations like incidents and hazards, a more efficient and safe traffic flow can be achieved.

Besides, once all the technical/scientific aspects related to the driverless car are resolved (which is in a good way), it is important also to have appropriate legislation for this kind of system[6] [Gasser et al., 2013] [NHTSA, 2013]. In fact, it is important to define clearly the responsibilities in the case of any accident (e.g., the fault could be attributed to the automotive manufacturer, the person inside the vehicle, the engineer, etc.). Different obstacles were already overcome in a few states in the USA (such as: Nevada, Florida or California) to allow, for instance, to the Google driverless car the testing of autonomous functionalities on public roads [GoogleCar, 2015]. Therefore, without clear legislation for driverless cars, these systems cannot be widely general-ized for our everyday life.

Nowadays, several interesting autonomous systems have been already deployed in different areas, for instance: in a warehouse as for KIVA®[7] system which is used notably by Amazon® warehouses to fulfill customer's orders in minimum time (cf.

---

[6]"The car to which we are accustomed and the way in which it is used are going to change radically and quickly," says William Clay Ford (the executive chair of the Ford Motor Company, founded by his great-grandfather, Henry Ford). He believes "the automatic car of the future will come sooner than we think. The main obstacle to its entry into service is the lack of a definitive legal framework. Laws need to be sufficiently robust to be enforceable in all jurisdictions around the world. Addressing this vital issue would pave the way for new players and stimulate the creativity of existing manufacturers."

[7]http://www.kivasystems.com, consulted January 2015.

(a) KIVA® system   (b) Autonomous UGV de- (c) Autonomous shuttles polyed by TEREX® com- (from Ultra Global PRT) in pany in ports   Heathrow airport, United Kingdom

FIGURE 1.6 (See color insert) Different actual deployed UGVs in closed environments.

Figure 1.6(a)), in a port as proposed by TEREX®[8] company (cf. Figure 1.6(b)) or airport shuttles as the one operational in London's Heathrow Airport (cf. Figure 1.6(c)).[9] Nevertheless, all these applications are performed in closed and mastered environments, in the sense that all the entities supposed inside this environment are known and their movements predicable. It is not at all the case in open[10] and highly cluttered environments as targeted by driverless cars with autonomous navigation function able to cope with any traffic conditions in rural, motorway and urban environments.

Since laboratories and innovative companies (like Google) propose a driverless car from scratch, the automotive manufacturers proceed with incremental (but with very reliable) developments using ADAS (Advanced Driver Assistance Systems). Several sophisticated ADAS have already entered the market, such as Automatic Parking, Adaptive Cruise Control, Lane Keeping Assistance or Collision avoidance system [ADAS, 2015]. This incremental (or bottom-up approach) is close in certain manner to what we claim in this manuscript though the use of multi-controller architectures (cf. section 1.4).

The road which leads to fully autonomous vehicles is still long and a lot of work remains to be done in the 3 phases cited above (perception-localization/decision/action) and on other different aspects (e.g., hybrid electric vehicles, battery and powertrain technology, etc.). Among the main challenging issues,

---

[8]http://www.terex.com/port-solutions/en/products/ automated-guided-vehicles, consulted January 2015.

[9]http://www.ultraglobalprt.com, consulted January 2015.

[10]The environment is not mastered in terms of the type of objects/persons which could be inside, or in terms of its sate conditions (like climatic one: winter, wind, fog, etc.).

and without exhaustivity (knowing the very important activity around these important topics), let us cite nevertheless some of them [Eskandarian, 2012] [Burns, 2013]:

1. In **general** (common for all the 3 phases):

    - Diagnostic analysis of the system to detect incoherence/risk in the perceptive or decision process as well as for the action.

    - Develop driver acceptance: using HMI (Human-Machine Interface) based for instance on Augmented Reality to trust the system.

    - Redundancy in terms of hardware and software (perception / localization / actuation) to reach a reliability of 10^{-9} (as much than for aeronautic systems).

    - Master the costs, maintaining a relative low cost of the obtained systems.

    - Increase the reliability and the security of communication exchanges, management/analysis of Big Data.

    - Enhance the modeling of: the driver (which could be inside other non-autonomous vehicles), the environment.

    - Develop more cooperative aspects (between vehicles themselves, between the vehicles and their infrastructures, or between the vehicle and the driver).

2. In term of **perception / localization**:

    - Enhance the sensors technologies (e.g., GPS, Camera, Laser, Radar, etc.) to better master the variability of car's environments (urban or rural, illumination of the scene, the rain, etc.).

    - Enhance the techniques of data fusion (sensors / map) to increase perceptive/localization accuracy and reliability.

    - Enhance the SLAM (Simultaneous Localization And Mapping) techniques.

    - Enhance the characterization/interpretation of the environment/scene (obstacles, pedestrian, other vehicles, etc.).

    - Enhance the modeling/characterization of the uncertainty.

    - Prediction of the trajectories of other mobiles/objects in the environment, occultations management, etc.

3. In term of **control (decision and action)**:

    - Develop more flexible and reliable strategies for autonomous navigation, which must deal with any new environment / situation. For instance, the Google car should know its exact environment (using an accurate map), and make at least 3 recognition of the area (using notably very dense sensor information) in order to permit a fully autonomous navigation.

- Enhance the developed control architectures which must deal with pre-dictable as well as non-predictable events, while mixing planning/re-planning and reactive behavior (cf. section 1.3.2).

- Develop appropriate control laws to minimize vehicle energy consumption.

- Enhance the modeling and the applied control laws / strategies to deal better with the uncertainty and the vehicle dynamic; the objective is to lead among others to safer and more comfortable vehicles behaviors, etc.

The different works developed in this manuscript deal mainly with the control aspects.

1.3   MAIN BACKGROUNDS AND PARADIGMS

This part of the chapter aims to emphasize the main concepts/paradigms/motivations/ definitions used throughout the rest of the manuscript. The objective is to introduce and clarify them in order to simplify the different explanations and developments given in the coming chapters.

1.3.1   Flexibility/stability/reliability definitions

Since we will use the terms Flexibility/Stability/Reliability to characterize the different proposals, let us give a short definition of what we mean by these words in the context of mobile robotics tasks:

- **Flexibility**: The ability of the autonomous robot to achieve the assigned task in several manners. The robot can, for instance, choose another path to reach its assigned final position.

- **Stability**: Used specifically to characterize the robot's control according to the Lyapunov definition (cf. Annex B, page 191). It is considered that if the robot's set-points are well defined (for instance, to achieve a sub-task such as obstacle avoidance) and that the control is proved stable, this means then that the sub-task will be achieved in a stable way.

- **Reliability**: The ability of the autonomous robot to achieve the assigned task even in the presence of different unmastered conditions such as the perception uncertainties or the task state conditions, etc.

1.3.2   Reactive *versus* cognitive control architectures

Reactive and cognitive control architectures for the navigation of mobile robots are notions which are highly used in the literature [Brooks, 1986] [Arkin, 1998] [Albus, 1991] [Adouane, 2005] [Eskandarian, 2012]. The widely used definition of reactive architecture corresponds to using only current sensor values (or values of a short time horizon) to decide the action to be achieved by the robot. This principle is well-known in the literature as stimuli-response robot behavior. To exhibit this principle, Braitenberg's machine (cf. Figure 1.7(a)) [Braitenberg, 1984] is a very significant example to show a purely reactive machine. A direct link exists between the sensor information and the actions of the robot's actuators. Figure 1.7(b) shows the links between the stimuli coming from the sensors and the actions of right and left wheels. In the case where for instance the left sensor receives more light than the right wheel will have a higher speed. This finally leads to the fact that the robot moves toward the light source (i.e., to the left in this example). Conversely, cognitive architectures, use much more data on the robot's environment and on its internal state to define the robot's actions.

Figure 1.8 gives a summary of these two control concepts [Arbib, 1981] [Arkin, 1998] [Adouane, 2005, Chapter 2]. According to the definitions given above and Figure 1.8, it is clear that the effective boundary between these two concepts (reactive and cognitive) is not as clear as what is supposed. The question is therefore, *"What is the limit so that an architecture could be characterized as reactive or cognitive?"*

We will not try to answer this quasi philosophical question, nevertheless, we will define in the following what we qualify as reactive control and what is considered more cognitive. Reactive control, as used in this manuscript, does not necessarily mean to strip completely the robot from any sophisticated perception or decision process. In fact, if it is possible to obtain online and reliable environment information and to react online to it, it would be obviously a pity not to use them. The main difference in what we call cognitive control is the fact to not use a complete environment knowledge to establish the robot's actions. Reactive control corresponds thus to reacting in real time to local environment knowledge without using any sophisticated task planning taking into account the overall environment knowledge. In addition, contrary to certain received ideas, the proposed reactive control architectures will be used while ensuring the stability and the reliability to achieve the assigned task (this aspect is particularly important to the targeted tasks such as transportation of persons). The limitation of the used reactive control is only related to the demonstration of the global optimality of the applied control/strategy to perform the overall assigned task. In fact, since the robot's actions are guided only by local perceptions, the global efficiency cannot be demonstrated in advance. Further, if the

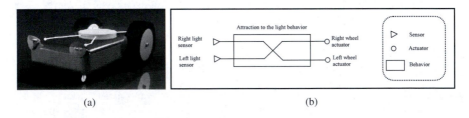

(a)   (b)

FIGURE 1.7   Braitenberg machine.

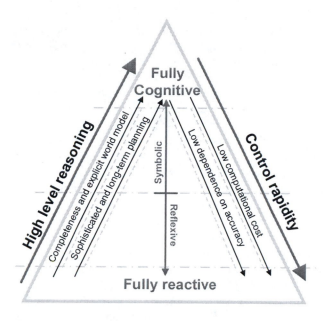

FIGURE 1.8 Reactive vs. cognitive control.

robot uses all the necessary environment knowledge to achieve its mission, the control is qualified as cognitive and it would be possible to prove the optimality of the taken decision/action.

The combination of reactive and cognitive approaches is an important functionality for an autonomous robot, to react for instance very quickly to unpredictable events (reactive mode) and to ensure the overall optimality of the task when the environment is better mastered (cognitive mode). Thus each mode can benefit from the other according to the navigation context. However, this Hybrid$_{RC}$ (Reactive/Cognitive) approach, raises the problem of knowing when it is better that the reactive functionality takes control and when it is better that the cognitive one take control. Chapter 4 will present a Hybrid$_{RC}$ control architecture.

1.3.3 Centralized *versus* decentralized control architectures for co-operative robotics

Chapter 6 will focus on the control of Multi-Robot Systems (MRS). The domain of cooperative robotics is an active research field and is currently linked to many key areas of application. The scientific issues associated with MRS concern formation analysis and control, cooperative perception, multi-robot localization, multi-robot task coordination, architectures for cooperation, and communication [Cao et al., 1997]. The coordination of a group of robots in a dynamic environment constitutes one of the fundamental problems (cf. chapter 6, page 139).

One of the key issues to fix before the development of MRS control architecture, corresponds to the possibility to centralize the control or to decentralize (distribute) it on the robotics entities [Cao et al., 1997] [Ota, 2006]. An architecture is called **centralized**, when a part or all of the sensory and/or decisional loops of each robotic entity is delocalized w.r.t. its physical structure, and managed by a central unit, called supervisor [Jones and Snyder, 2001], or central planner [Noreils, 1993] [Causse and Pampagnin, 1995]. A centralized architecture is usually a synonym of Top-Down approach and can be imaged by a conductor (the supervisor) that directs his musicians (mobile robots). These architectures imply a global knowledge of each element of the system and they require high computational power, massive information flow and they are generally not robust (due notably to the dependence on a single controller/supervisor). In contrast, in a **decentralized (distributed)** control approach, each element of the system has its own perceptions / decisional process. This kind of control implies a reduced number of communicated signals and data knowledge. In fact, each robotic entity does not need to have the overall environment knowledge before acting on its environment. Decentralized control, if well mastered, is then more flexible to deal with MRS having a large number of entities and is generally a synonym of Bottom-Up approach [Adouane, 2005, chapter 2].

The possibility also exists to centralize only a part of the control and let the other part be decentralized (hybrid (centralized/decentralized) control) [Fukuda et al., 2000]. The centralized control is applied to determine the general strategies and tasks to be performed by the MRS, and the decentralized part takes over for the navigation and local actions. It is important also to link the notion of a centralized/decentralized control architecture and cognitive/reactive architecture where a reactive robot reacts generally to its local perceptions. This observation permits us to say that if reactive robots evolve in MRS, they will be mainly controlled in a decentralized manner.

In our developed control architectures related to MRS (cf. chapter 6), decentralization of the control is always favored, but in certain situations where the environment (or the task) permits it, global information on the system could be used to enhance the MRS control.

1.3.4 Boundary limit between planning and control

To perform any elementary robot's behavior (obstacle avoidance or target reaching, for instance), the robot has to follow, generally, these two steps:

1. Define appropriate set-points (e.g., according to planned path for instance) which takes into account: the current robot's and environment's state, and also the sub-task to achieve. These set-points could be obtained for instantaneous, short- or long-term robot evolution. Indeed, for different situations, the robot has for instance to re-plane its trajectory or to define another target to reach (cf. section 2.4, page 37).

2. Once the set-points are obtained, an appropriate control law must be used to attain the assigned set-points.

Both of these steps are important to achieve safely and in a reliable way the assigned robot's task. For the first step (set-points planning), a part of literature does not take into account the robot physical constraints (such as its non-holonomy or its maximal velocity) or the uncertainty related to the perceptive or to the modeling aspects (robot / environment, etc.). Artificial Potential Field (APF) [Khatib, 1986], Voronoï diagrams and Visibility graphs [Latombe, 1991], Navigation Functions [Rimon and Daniel.Koditschek, 1992] or planning-based Grid checking and trajectory generation [Pivtoraiko and Kelly, 2009] are among these road-map-based methods. Even if all these methods are very interesting, they may not be reliable for the robot, since it could not always follows the assigned original planning.

The other part of literature considers that to obtain accurate, flexible and reliable robot control, it is essential to take into account the physical constraints and the model uncertainties from the beginning [Choset et al., 2005] [Kuwata et al., 2008]. The rate of success to control effective robots is obviously more important when using this latter approach. Nevertheless, it increases considerably the complexity to obtain the set-points and is highly time-consuming.

In this manuscript, to preform reliable navigation (for mono as well as for multi-robot systems), the actions on the set-points as well as on the control laws will be done and motivated in all the following chapters. It is aimed therefore in the presented works to really separate between what is sent to the controllers as set-points and the effective robot evolution (which depends on, among other things, the robot model and constraints, the environment conditions, etc.).

1.4   FROM BEHAVIORAL TO MULTI-CONTROLLER ARCHITECTURES

This section will start by giving the genesis of using multi-controller architectures and the most important features to master before creating an actual reliable and efficient tool to control autonomous mobile robots (cf. subsection 1.4.2).

There exist in animals innate behaviors which could be qualified as atomic (or elementary) in the sense that they are not reducible to simpler behaviors (directly observable). In general, all animals' motor actions (coordination of a set of muscle activities) are included in this category. These behaviors are the building blocks with which the behavior of a higher level can be built and described [Arbib, 1981] [Anderson and Donath, 1990]. Further, autonomous mobile robots can have to perform several tasks (cf. section 2.1), for instance going to a specific target (location) in the environment while avoiding obstacles and in certain cases while maintaining a formation (as targeted notably in our works (cf. chapter 6)) and so on. In addition, these sub-tasks must also be achieved generally while guaranteeing multi-objective criteria to obtain for instance reliable and smooth robot navigation. All these sub-tasks and several criteria increase considerably the complexity to attain efficient autonomous robot navigation.

To address this complexity (in terms of task definition and multi-objective criteria), the control architectures can be elaborated in a modular and bottom-up way as introduced in [Brooks, 1986] and so-called behavioral architectures [Arkin, 1998].

Behavioral control architectures[11] are based on the concept that a robot can achieve a global complex task while using only the coordination of several elementary behaviors [Arbib, 1981] [Brooks, 1986] [Anderson and Donath, 1990]. To tackle this complexity, behavioral control architecture decomposes the global control into a set of elementary behaviors/controllers (e.g., attraction to a target, obstacle avoidance, trajectory following, etc.) to better master the overall robot behavior. Indeed, each behavior can be tested either individually or collectively with other behaviors. The goal is to verify the reliability and the efficiency of the corresponding behavior to achieve a determined sub-task.

Nevertheless, several challenges remain to be addressed before obtaining an effective and reliable multi-controller[12] architecture. Among the main objectives of our works is to lead to stable and reliable multi-controller architectures while maintaining a high level of flexibility, necessary to tend toward fully autonomous vehicles (cf. section 1.2).

The goal of any control architecture is to reach and/or maintain desired system states (configurations). In the framework of multi-controller architectures, this takes place through simultaneous control of several levels of action and decision (cf. Figure 1.9). They could be summarized as follows:

- **Level 1** corresponds to all intervening elements for the management and hardware execution of the control set-points sent to the robot's wheels, and it deals therefore with the robot's actuators, inertia, or in general with its actual dynamics. This level is responsible to ensure that the control commands (frequency, values, etc.) are compatible with the robot's physics/dynamics.

- **Level 2** covers the inner workings of each elementary controller (sub-task), and corresponds to the used methodology to create a correaltion between the robot's perception and what it must follow as set-points to achieve the sub-task. Levels 1 and 2 are therefore closely linked in the sense that to have a reliable controller, it is important to well define appropriate set-points as well as control laws (cf. subsection 1.3.4).

- **Level 3** corresponds to how to coordinate the activation/action of the multitude of controllers embedded in the same multi-controller architecture (cf. subsection 1.4.1).

- **Level 4** corresponds to the process of coordination between a set of controllers' aggregates, each aggregate permitting the achievement of a complex task. This level corresponds more to the framework of artificial life where a creature (the robot) must live (perform autonomously several complex tasks) and interact in a complex environment.

---

[11] Which are obtained by the aggregation of several elementary behaviors cohabiting in the same structure. The author in [Arkin, 1998] claims that "Behavior-based roboticists argue that there is much that can be gained for robotics through the study of neurosciences, psychology, and ethology."

[12] The term multi-controller will replace in what follows, the term behavioral because it has been wildly investigated in our works the use of automatic control theory to confirm among others the reliability of each controller as well as the overall multi-controller architecture (cf. subsection 1.4.2).

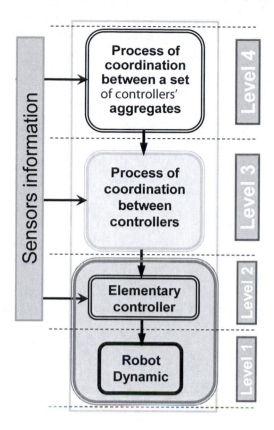

FIGURE 1.9 Different levels of decision/action in multi-controller architectures.

To obtain fully autonomous vehicles, the 4 levels (given above) should be mastered. Levels 1 to 3 will be largely addressed in the following chapters. The next subsection will detail the different existing techniques for the coordination of controllers (Level 3).

1.4.1 Multi-controller coordination

The development, the mastery, and the working coherence of a multi-controller architecture pass inevitably by the mastering of the flux of commands generated by the multitude of behaviors/controllers cohabiting in the same control structure. In other words, it is necessary to determine among the set of the generated commands by the elementary behaviors at every sample time, those which are going to contribute effectively to the future actions of the robot. The existing interactions between controllers must be by consequent completely mastered. Indeed, it is the mastery of behavior relations that will allow us to get more and more complex behaviors (via the addition

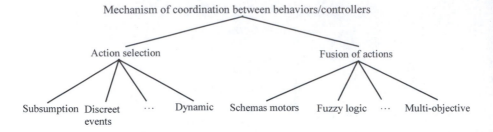

FIGURE 1.10 Tree of the different mechanisms of coordination between behaviors/controllers.

of new controllers) without losing the flexibility, the robustness and the predictability of the proposed control architectures. The architectures proposed for instance in [Maes, 1991] or in [Parker, 1998], give an outline of some mechanisms of behavior coordination which can cause some cyclic, blocking and even unpredictable robot situations. This is essentially due to the non-mastery of the existing interactions between the elementary behaviors.

In multi-controller architectures as investigated in this manuscript, two major principles of controller/behavior coordination exist: *action selection* and *fusion of actions*, which lead, respectively, to competitive or cooperative control architectures. Figure 1.10 shows a tree representing the different mechanisms of coordination [Pirjanian and Mataric, 2001] [Adouane, 2005].

In competitive architectures (*action selection*), the set-points sent to the robot's actuators at each sample time are given by a unique controller which is selected among a set of possible controllers. The principle of competition can be defined by a set of fixed priorities like in the subsumption architecture [Brooks, 1986] where a hierarchy is defined among the controllers. The resulting actuators commands are given by the active controller with the highest level of priority. The *action selection* can also be dynamic without any hierarchy between behaviors [Maes, 1989] [Mataric et al., 1995]. All the controllers can mutually activate or inhibit each other. At each sample time, the selected controller is the one with the highest level of activation. This level of activation is generally computed according to a linear combination of a number of internal and external stimuli. Among the different competitive architectures, let us cite [Maes, 1989] [Mataric et al., 1995] with the Dynamical Action selection architecture, or [Drogoul, 1993] with the EthoModeling Framework (EMF) which is inspired from the working mechanism of social insects in order to build reactive agents. Parker in [Parker, 1999] proposes the ALLIANCE architecture with an adaptive action selection mechanism to achieve cooperative missions with a team of mobile robots. The robots select an appropriate action based on the requirements of the mission, the activities of other robots, the current environmental conditions and finally on their own internal states.

In cooperative architectures (*fusion of actions*), the set-points sent to the robot's actuators are the result of a compromise or a fusion between controls generated by several active behaviors. These mechanisms include fuzzy control [Saffiotti et al., 1993] [Arrúe et al., 1997] [Wang and Liua, 2008] via the process of defuzzification, or the multi-objective techniques to merge the controls [Pirjanian, 2000]. Among these cooperative architectures, the schema-based principle [Arkin, 1989b] is among the ones which have an important impact in the scientific community. This architecture uses in general the same potential field technique [Khatib, 1986], [Arkin, 1998] for the computation of the response of each elementary controller and also to encode the robot's behavioral response. There exist also in literature several other proposed mechanisms of coordination like those used in [Connell, 1990], [Mataric, 1992], [Ferrell, 1995], [Sigaud and Gérard, 2000], [Simonin, 2001], [Adouane and Le Fort-Piat, 2004], [Dafflon et al., 2015], etc.

Although the *fusions of actions* process gives a very interesting robot behaviors, the stability of the overall control architecture is generally very hard or impossible to demonstrate. At the contrary, overall stability of control architectures based on the *action selection* process are usually much easier to demonstrate even when a switch between the controllers occurs [Adouane, 2009a] [Benzerrouk et al., 2009] (cf. chapter 3).

1.4.2   Multi-controller architectures (main challenges)

As emphasized above, the main advantages of multi-controller architectures arise from their Bottom-Up construction and flexibility to deal with several complex tasks. Nevertheless the lack of possible analytic analysis (e.g., stability or robustness) of the obtained architecture reduces drastically the possible use of this paradigm to perform tasks needing a high level of reliability and safety. This is notably the case of autonomous transportation or service robotics tasks (where close interaction between the robots and humans exists).

The challenging issue of this kind of architecture is therefore to have the possibility to prove its reliability while introducing mathematical analysis for each developed controller, as well as for the overall multi-controller architecture. The analysis of the overall architecture means, among other things, to master the coordination (while avoiding at maximum jerking in the control commands (smooth control)) of the multitude of controllers characterizing such architecture. At this aim, it is considered in some studies to investigate the potentialities of hybrid systems[13] controllers [Zefran and Burdick, 1998] to provide a formal framework to demonstrate the robustness and the stability of multi-controller architectures [Adouane, 2009a] [Benzerrouk et al., 2009] (cf. chapter 3). In their simplest description, hybrid systems are dynamical systems modeled as a finite-state automaton. These states correspond to a continuous dynamic evolution, and the transitions can be enabled by particular conditions reached by the continuous part. This formalism thus permits a rigorous automatic control analysis of the overall architecture [Branicky, 1998].

Thus, the main challenge inherent in multi-controller architectures is to demonstrate their overall stability. In fact, it is not enough to demonstrate the stabil-

---

[13]Which allows controlling continuous systems in the presence of discrete events.

ity of each elementary controller to guarantee the overall stability of the multi-controller architecture, mainly if the switches between controllers can occur arbitrarily [Branicky, 1993]. To highlight this, let us present the example given in [Liberzon, 2003] where the switch between two systems 1 and 2 is considered. The global system state is $x = (x_1, x_2) \in \mathbb{R}^2$, the switch occurs thus in the plan. Further, while supposing that the two elementary systems are stable, each system applied separately leads then to an equilibrium state x as shown in Figures 1.11(a) and 1.11(b). However, switching between the two systems does not necessarily lead to stability. Figure 1.11(c) shows that according to the different switching moments, the overall systems could be unstable [Branicky, 1993]. Authors in [Johansson et al., 1999] modeled a hybrid system using an automata approach, where each node corresponds to a dedicated control law. The authors highlighted that hard switches between controllers may lead to the Zeno phenomenon which exhibits an infinite number of discrete transitions in finite time. This phenomenon appears when the robot state is always in the boundary-limit where the discrete event (activating the switch) becomes true. The author in [Egerstedt, 2000] has resolved this problem while adding another node (control law) in the automaton.

To summarize, before taking full advantage of the potentialities of multi-controller architectures, some challenging issues should be addressed:

- For **elementary controllers**, they must be stable and reliable for different environment contexts (e.g., cluttered or not, dynamic or not, etc.).

- For the **coordination mechanism between controllers**:

    - Master the coordination between controllers' actions (hard switch or merging) in order to achieve safely and efficiently the assigned task,

    - Avoid jerking/discontinuities in term of the robot's control. The objective is to obtain stable and smooth switching between controllers.

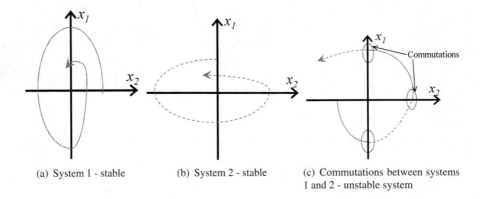

(a) System 1 - stable   (b) System 2 - stable   (c) Commutations between systems 1 and 2 - unstable system

FIGURE 1.11 Hybrid system commutations which can lead to an unstable overall system.

To respond to the items given above, several ideas have been developed in this manuscript consisting summarily of:

- Use intensively automatic control specifications/characteristics to prove the features of the proposed elementary controllers (e.g., obstacle avoidance, target reaching/tracking, etc.).

- Use hybrid systems theory as a formal framework for controller design and coordination (cf. chapter 3).

1.5 NAVIGATION BASED ON TRAJECTORY OR TARGET SET-POINTS

Different strategies of autonomous navigation have been proposed in the literature [Lee and Litkouhi, 2012] [Gu and Dolan, 2012]. The most popular approaches are based on following a pre-defined reference trajectory (time-parametrized) [Bonfè et al., 2012] [Kanayama et al., 1990]. These methods link the robot control to this reference trajectory which could be defined by a combination of path planning and trajectory generation techniques [LaValle, 2006].

Typically to obtain the reference path to be followed by the robot, arc-lines, B-splines or polynomial equations are used over points [Connors and Elkaim, 2007] [Horst and Barbera, 2006] [Lee and Litkouhi, 2012]. In [Gu and Dolan, 2012] a feasible path is obtained using a polynomial curvature spiral. In [Bonfè et al., 2012], the trajectory generation method provides a smooth path considering the kinodynamic constraints of the vehicle. In [Labakhua et al., 2008], trajectories are built using user-assigned points and interpolation functions such as cubic splines, trigonometric splines and clothoids. Moreover, velocity profiles along the trajectory are specified to improve the passengers' comfort which is related to the vehicle's acceleration. Nevertheless, trajectory generation presents some drawbacks, such as the necessity of a specific planning method, the proof of guarantee of continuity between different segments of the trajectory or the complexity for replanning (mainly in dynamic and uncertain environments).

Furthermore, since almost all of the navigation strategies are based on path following or trajectory tracking [Laumond, 2001], most of the control laws proposed in the literature deal with these set-points. Hence, several dedicated control methods have been proposed in the literature [Blazic, 2012] [Fang et al., 2005] [Goerzen et al., 2010] and [Pesterev, 2012]. In [Kanayama et al., 1990] [Blazic, 2012] and [Vilca et al., 2013a], nonlinear control laws for trajectory tracking are synthesized for a unicycle robot using Lyapunov stability analysis. A trajectory tracking control for a farm vehicle, incorporating sliding in the kinematic model, is proposed in [Fang et al., 2005]. For the path-following problem, a control law for a tricycle robot is proposed in [Pesterev, 2012] and [Samson, 1995]. They are based on feedback linearization and chained form representation [Laumond, 2001]. The path-following controller allows making the lateral and longitudinal control of the vehicle independent along the reference trajectory. Further, the path-following controller allows smoother convergence to the desired path than the trajectory tracking controller [Siciliano and Khatib, 2008]. The trajectory tracking controller allows tracking the trajectory with a desired velocity profile, while the path-following controller acts only on the orientation to drive it along the path. Both, the path-following and trajectory tracking controllers require the pose of the closest point to the trajectory (w.r.t. robot configuration) and/or the value of curvature at this point, at each sample time [Laumond, 2001] [Blazic, 2012]. Although there exist a multitude of techniques to compute these parameters, they can add an error in certain situations thereby influencing negatively the mobile robot control [Siciliano and Khatib, 2008] [Fang et al., 2005].

Besides, contrary to following or tracking a trajectory to lead the robot toward its objective, few works in the literature propose to use only specific waypoints in the environment to lead the robot toward its final objective. In [Aicardi et al., 1995], the authors propose a navigation via assigned static points for a unicycle robot. Nevertheless, the definition of the mission is less accurate because this strategy does not consider the kinematic constraints of the robot (maximum velocity and steering), the orientation error and the velocity profile of the robot when it reaches the assigned point. Harmonic Potential Field (HPF) is used in [Masoud, 2012] to guide a UAV to a global waypoint with a position and a direction of arrival. The author proposes a virtual velocity field which allows us to consider the UAV physical model. Each vector component of the field is treated as an intermediate waypoint with which the robot must comply in order to reach the global waypoint. Nonetheless, HPF requires a complex mathematical modeling for different shapes or dimensions of the obstacles.

In general, it will be shown all along the manuscript the different motivations related to the use of only targets reaching/tracking behavior to perform several robot sub-tasks (cf. section 2.4, page 37). For instance, chapter 5 will highlight the fact that only a few waypoints (static or dynamic), appropriately positioned in the environment, are sufficient to guarantee safe vehicle navigation. The main advantage to proceed like this is to enhance the flexibility of the robot's movements, since it is allowed to have more maneuvers between waypoints. Chapter 6 will show also the different advantages of using targets reaching/tracking controller to define/perform the navigation in formation of a group of robots.

1.6 CONCLUSION

This chapter introduced briefly the domain of **autonomous mobile robotics** while focusing on **ground robots** with its main achievements/developments/levels of autonomy and challenges. It permitted us also to emphasize/clarify the main backgrounds/paradigms/motivations and definitions useful for the rest of the manuscript.

The notions of **flexibility**, **stability** and **reliability** have been defined in the context of the developed **control architectures**. In term of robotics paradigms, the concept of **reactive control** architecture was clarified. In fact, the main difference with **cognitive/deliberative control** is the use or not of complete environment knowledge to establish the robot's actions. Reactive control corresponds thus to **real-time** response to local environment knowledge without using any sophisticated **task planning** (taking into account the overall environment knowledge). In addition, contrary to certain received wisdom, the proposed reactive control architectures will be used while ensuring stability and reliability to achieve the assigned task (this aspect is particularly important according to our privileged targeted tasks such as **transportation of persons**). In addition, we also defined/motivated the challenging aspects related to **decentralized approaches** (w.r.t. **centralized** ones) for the **control of multi-robot systems**; boundary limits between **planning** and control; and as far as possible to use a navigation based on **target set-points** instead of **trajectory following/tracking**. We also emphasized the genesis and the potentialities of **multi-controller architectures** and their main issues to resolve in order to become a very efficient tool to control mobile robots.

It is to be noted that the term "robot," used in the rest of the manuscript, will designate mainly grounded mobile robots with wheels. The concepts (control architectures / strategies of navigation) developed in this manuscript have as a main objective to be **generic** enough to be used for different robot structures (unicycle, tricycle, etc.) and therefore the term mobile robot or vehicle will be replaced (almost everywhere in the manuscript) as above by the term robot (or mobile robot). The focus will be made in this manuscript on the control of robots (mono or multi-robot entities) to perform **autonomous navigation** in **highly cluttered and dynamic environment** (cf. section 2.1, page 26). In the adopted methodology, it is taken as principle that if the proposed control architecture can be reliable in that kind of complex and constrained environment, it could be obviously even more reliable in less constrained ones.

CHAPTER 2

Autonomous navigation in cluttered environments

CONTENTS

THIS CHAPTER introduces important elementary building blocks character-izing the different multi-controller architectures developed throughout this manuscript. Among these blocks let us cite: obstacle avoidance (based on Limit-Cycles) and target reaching/tracking elementary controllers. Their set-point definitions and main features will be emphasized showing their general use in several navigation tasks. Furthermore, a complete multi-controller architecture, dedicated to reactive navigation in cluttered environments will be presented. Some details about the perceptive aspects will also be discussed.

2.1   OVERALL NAVIGATION FRAMEWORK DEFINITION

Autonomous mobile robot navigation (cf. Figure 2.1) is a complex problem of major interest to the research and industrial communities (cf. section 1.1, page 2). Systems capable of performing efficient and robust autonomous navigation are unquestionably useful in many robotic applications such as manufacturing technologies [Sezen, 2011], urban transportation [Vilca et al., 2015a], assistance to disabled or elderly people [Martins et al., 2012] and surveillance [Stoeter et al., 2002]. Although much progress has been made, some specific technologies have to be improved for widespread use in actual environments (cf. section 1.2, page 7). Further, to perform a fully autonomous robot navigation, in addition to having accurate perception and localization capacities [Thrun et al., 2005] [Choset et al., 2005] [Siegwart et al., 2011] (which are not the focus of the manuscript), the robot must have the ability to be controlled online in different kinds of environments (e.g., cluttered or not, dynamic or not, uncertain or not, etc.) and to react safely to unpredictable events. Thus, the used control architecture must permit us to answer this important question "How do we reach safely and efficiently a predetermined location in an environment while taking into account available environment knowledge (the road limits for instance) and reacting online to unpredictable events (e.g., other robots, obstacles, etc.)?"

Furthermore, it is not sufficient to guarantee only the reliability and the safety of the navigation; the robot must also insure, in transportation applications for instance

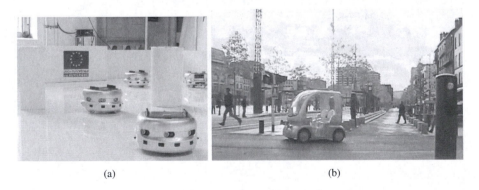

(a)   (b)

FIGURE 2.1   (See color insert) Autonomous navigation of (a) a group of mobile robots (Khepera®) and (b) an electric vehicle (Cycab®) in an urban environment (Clermont-Ferrand, France). MobiVIP project (Predit 3).

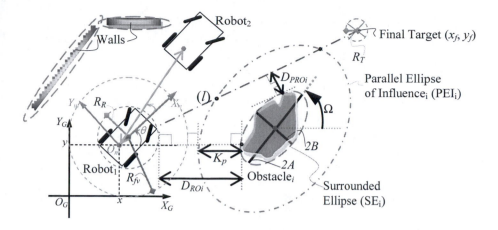

FIGURE 2.2   Robot's pose and its perceptions for mainly reactive navigation.

[Fleury et al., 1993] [Vilca et al., 2013b], smooth navigation for the comfort of the passengers. In [Gulati, 2011], the author characterizes this smooth navigation while using a cost function which reflects the trade-off between the travel time and the integral of acceleration (which characterizes the jerking amount of angular and linear robot velocities). Fully autonomous navigation needs therefore to satisfy simultaneously a multitude of criteria. For this aim it is important to have a reliable, safe and flexible control architecture [Minguez et al., 2008]. Different navigation strategies (using dedicated control architectures) have been proposed in the literature. They permit autonomous navigation even in dynamic and cluttered environments. This means that the obstacle avoidance function is always an important primitive and is tightly inherent to the performed autonomous navigation strategy. Therefore, special attention should be taken for its development [Minguez et al., 2008] (cf. section 2.2).

Before giving more details about the different control architectures and their inherent obstacle avoidance strategies let us describe in what follows the adopted overall navigation framework, which has been chosen in order to be generic enough to depict a large number of autonomous navigation tasks. Autonomous robot navigation aims, in the proposed generic framework, to lead the robot from its initial configuration, to a final configuration (called Final Target), avoiding any obstacle (which could have different shapes, cf. Figure 2.2) and in certain cases, the robot has to maintain a certain formation (distance / orientation) with other navigating robots (chapter 6 main focus is multi-robot navigation in formation). The robot navigation could be done even with reactive control (while acting online according to the robot's local perception, or with cognitive control (while following an already planned trajectory / waypoints set-points; cf. chapters 4 and 5 for more details about the proposed techniques). The desired robot movement needs to be safe and smooth along all its displacement. One supposes in the setup that the robot and the final target to reach are surrounded by circle shapes with a radius R_R and R_T respectively (cf. Figure 2.2). In addition, the robot's maximum field of view is considered as a circle centered on the robot with a radius R_{fv}. This circle corresponds to the maximum distance where the robot can detect any other object (obstacles, other robots, etc.). For obstacles/walls, it is supposed that they can be surrounded by appropriate ellipses (cf. Figure 2.2), given by equation 2.1. The choice of an ellipse shape rather than a circle, as used in several works (for instance in [Kim and Kim, 2003], [Jie et al., 2006] or [Adouane, 2009b]), is to have one more generic and flexible means to surround and fit accurately different obstacles shapes.

Among the examples of shapes which can be appropriately fitted with an ellipse and less by a circle is a wall (or in general, any longitudinal or thin shapes). Figure 2.3 shows this kind of configuration. In fact, if we would like to surround the wall given in this figure by an appropriate circle, this one will have a large radius which induces more robot path distance to avoid it safely [Kim and Kim, 2003] (cf. Figure 2.3(a)). Figure 2.3(b) shows that the ellipse better fits the shape of this wall. This figure shows also uncertain perceptions obtained by infrared sensors on one side of the wall (left side). Several examples of robot navigation in cluttered environments will be shown notably in sections 2.5.5 and 4.4.4 (page 92). These simulations will highlight, among other things, the validity and the relevance to surround different obstacles/walls/sidewalk/etc. by appropriate ellipses shapes. An ellipse is defined in what follows by:

$$a(x - h)^2 + b(y - k)^2 + c(x - h)(y - k) = 1 \qquad (2.1)$$

with:

- h, $k \in \mathbb{R}$, are the coordinates of the ellipse center,

- $a \in \mathbb{R}^+$, is related to the half-length $A = 1/\sqrt{a}$ of the ellipse's longer side (major axis),

- $b \in \mathbb{R}^+$, is related to the half-length $B = 1/\sqrt{b}$ of the ellipse's shorter side (minor axis), thus $b \geq a$,

- $c \in \mathbb{R}$, is related to the ellipse orientation (if $a \neq b$) $\Omega = 0.5atan(c/(b - a))$ (cf. Figure 2.2). When $a = b$ equation 2.1 becomes a circle equation (Ω will not give thus any more information).

The surrounded ellipse parameters (h, k, A, B and Ω) (cf. equation 2.1 and Figure 2.2) can be obtained online as will be emphasized in section 2.5.2.

2.2 SAFE OBSTACLE AVOIDANCE AS AN IMPORTANT COMPONENT FOR AUTONOMOUS NAVIGATION

An interesting overview of obstacle avoidance methods is accurately provided in [Minguez et al., 2008]. Obstacle avoidance behavior is tightly linked to the used control architecture class. This latter can be split into two categories (cognitive and reactive (cf. section 1.3.2, page 11)):

The first, categorized as cognitive (or deliberative), makes its main focus on

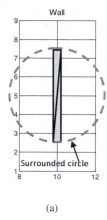
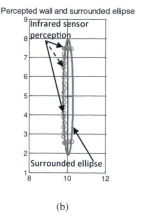

(a)   (b)

FIGURE 2.3 Different shapes to surround and to avoid safely obstacles; (a) and (b) interpolated wall using, respectively, a circle and an ellipse shape.

the trajectory/path planning and re-planning [Pivtoraiko and Kelly, 2009], while generally taking into account the overall environment knowledge. It is to be noted that the terms trajectory or path are used, respectively, if the time is taken or not into account during the planning phase. For simplification, in what follows, the term trajectory will be used to express the two cases. The obtained trajectory takes into account all obstacle configurations (and maybe its dynamic) in the planning step. In fully cognitive navigation, once a trajectory is obtained, the robot follows it as accurately as possible using the dedicated control law, for instance using the well-known control laws proposed in [Kanayama et al., 1990] or [Samson, 1995]. Among the multitude of methods linked to cognitive architectures we can cite Voronoï diagrams and visibility graphs [Latombe, 1991]; navigation functions [Rimon and Daniel.Koditschek, 1992] or planning-based grid map [Choset et al., 2005]; Rapidly exploring Random Tree (RRT) [Lavalle, 1998], and Sparse A* Search (SAS) [Szczerba et al., 2000]. Widely used in cognitive control architectures are pre-planned reference trajectories, which means that they are properly selected before robot movement [Morin and Samson, 2009]. A large class of roadmap-based techniques [Szczerba et al., 2000] use optimization to choose between a set of admissible trajectories [Lavalle, 1998] [Brock and Khatib, 1999] [Ogren and Leonard, 2005]. In these methods, it is possible to deal with a changing environment while regularly re-planning the robot's trajectory [Fraichard, 1999] [Van den Berg and Overmars, 2005]. However, planning and re-planning require significant computational time to deal with the algorithmic complexity. Therefore, in real motion conditions where the environment could be very cluttered, uncertain and/or highly dynamic, these methods may not be reliable, due among other things, to the large amount of necessary time to obtain the new planned trajectory [Minguez et al., 2008] [Parker, 2009].

The second part of the literature, categorized as reactive, considers that the robot needs to answer in real time to its current perceptions [Brooks, 1986] [Arkin, 1989b] [Adouane and Le-Fort-Piat, 2004] and needs thus less knowledge about the overall environment. Local sensor information is used rather than a prior important knowledge on the environment [Egerstedt and Hu, 2002] [Toibero et al., 2007] [Adouane, 2009a]. The robot reacts therefore with stimuli-response behavior (generally bio-inspired [Arbib, 1981]) and does not need any important planning process to achieve the navigation. In fully reactive navigation, at each sample time, the robot should follow a defined set-points, according to its local perceptions and current objectives (for instance, reach a pre-defined location). The obstacles / walls / pedestrians / etc. are thus discovered and avoided in real time. In [Khatib, 1986] the author proposes a real time obstacle avoidance approach based on the principle of artificial potential fields. He assumes that the robot's actions are guided by the sum of attractive and repulsive fields. In [Arkin, 1989b] the author extends Khatib's approach while proposing specific motor schema for mobile robot navigation. Another interesting approach, based on a reflex behavior reaction, uses the Deformable Virtual Zone (DVZ) concept, in which the robot's movement depends on a risk zone surrounding the robot [Zapata et al., 2004]. If an obstacle is detected, it will deform its DVZ and the approach consists of minimizing this deformation by modifying the control vector. This method deals with any obstacle shape, however, it suffers as schema motors from local minima problem. In general, this school of thought (the reactive one) does not require high computational complexity since the robot's actions must be given in real time according to local perception [Arkin, 1998]. Obviously while taking this philosophy for robot navigation, the overall robot's movement cannot be considered as optimal, mainly if the environment is complex with a multitude of obstacles and maybe trapped regions [Ordonez et al., 2008] (cf. section 1.3.2, page 11).

Besides the two existing schools of thought (reactive and cognitive) multi-controller architectures (cf. section 1.4, page 15) could be used for both kinds of architectures.[1] In fact, the multi-controller architecture's main feature permits us to isolate each of its components (either low or high level) and to make it as reliable as possible before incorporating it in the overall control architecture. The obstacle avoidance is one of the most important building blocks to achieve autonomous navigation of mobile robots in cluttered environments. It will be accurately detailed in section 2.3 before integrating it, firstly in simple multi-controller architectures (cf. section 2.5) and later in the manuscript, in more sophisticated multi-controller architectures. In fact, several enhancement/developments of the basic control architecture will be highlighted; for instance, the stability of the overall control architecture in chapter 3; the integration of the cognitive possibilities in chapters 4 and 5 and finally the extension to multi-robot systems in chapter 6.

---

[1]Chapter 4 will focus on a control architecture which exhibits reactive and cognitive skills.

2.3 OBSTACLE AVOIDANCE BASED ON PARALLEL ELLIP-TIC LIMIT-CYCLE (PELC)

In what follows we present generic component, called Parallel Elliptic Limit-Cycle (PELC), used as an elementary safe trajectory, as much for reactive than for cognitive navigation. The PELC is based on a generic mathematical formulation of a limit-cycle (cf. subsection 2.3.1). Subsection 2.3.2 will present an appropriate reference frame useful to achieve different sub-tasks (obstacle avoidance is one among these sub-tasks).

2.3.1 Elementary PELC

Limit-cycles correspond, as they will be used in what follows, to specific trajec-tories which have always a fixed orbit where converge all the trajectories starting inside or outside (the given orbit). The methodologies based on limit-cycles have been used in the literature to perform intuitive and efficient obstacle avoidance behavior [Kim and Kim, 2003] [Jie et al., 2006] [Adouane, 2008] [Adouane, 2009b] [Soltan et al., 2011]. They are defined according to a circular [Adouane, 2009b] or an elliptic [Adouane et al., 2011] periodic orbit. These periodic orbits can guarantee, if they are well-dimensioned (far enough from any obstacle) and accurately followed, avoidance of any obstructing obstacle. Unlike potential field methods [Khatib, 1986], only the most disturbing obstacle impacts the robot's trajectory, which permit us to avoid local minima and oscillations. The modeling of limit-cycles are very close to those based on Harmonic Potential Fields (HPF), and this approach takes inspira-tion from the description of the dynamic movement of fluids around impenetrable obstacles [Masoud, 2012]; Dynamical System approaches (DS, by dynamical sys-tems it is meant a coupled set of "n" autonomous first-order differential equations) [Khansari-Zadeh and Billard, 2012]. These two well-known methodologies permit, while following their defined trajectories (according to HPF or DS set-points), to safely avoid obstacles by bypassing them. Nevertheless, HPF as well as DS require very complex mathematical modeling to avoid any obstacle's shape. Contrary to that, the defined generic PELC does not need any complex computation, from the moment that the parameters of the surrounded ellipse are obtained $((h, k), A, B, \Omega$, cf. equa-tion 2.1 and Figure 2.2). In addition to obstacle avoidance, the proposed PELC can be easily used for several robot navigation sub-tasks (cf. section 2.4). Therefore, it constitutes a uniform component to perform safe and reliable navigation.

The work given in [Adouane et al., 2011] has permitted us to extend possible Cir-cular Orbital Limit-Cycles (COLC) to Elliptic Orbital Limit-Cycles (EOLC), where COLC is only a particular case of EOLC when the major axis is equal to the minor axis. In what follows, an even more generic formulation of limit-cycle trajectories is presented, permitting us to obtain an orbital Parallel Elliptic Limit-Cycle (PELC). This PELC has a constant offset according to the surrounded ellipse (cf. Figure 2.2). Before we give more details about PELC, let us give a short definition of the gen-eral principle of parallel curves, frequently called for mathematical definition "offset curves" [Segundo and Sendra, 2005]. These curves are obtained from a base curve by a constant offset, either positive or negative, in the direction of the curve's normal. The two branches of the parallel curve at a distance K_p away from a parametric represented base curve $(f(t), g(t))$ are given by:

$$x = f \pm \frac{K_p \dot{g}}{\sqrt{\dot{f}^2 + \dot{g}^2}}; y = g \mp \frac{K_p \dot{f}}{\sqrt{\dot{f}^2 + \dot{g}^2}}; \qquad (2.2)$$

where $\dot{f} = df/dt$ and $\dot{g} = dg/dt$. It is important to mention that a parallel curve to an ellipse is not an ellipse (but an equation of 8 order), except obviously if the offset $K_p = 0$.

The differential equations of a PELC are given as follows (cf. Figure 2.4):

$$\dot{x}_s = ry_s + \mu x_s(1 - \Psi)$$
$$\dot{y}_s = -rx_s + \mu y_s(1 - \Psi) \qquad (2.3)$$

where:

- $\Psi = [4(z_1^2 + 3z_2)(z_2^2 + 3z_1z_3) - (z_1z_2)^2 + 18z_1z_2z_3]/(9z_3)^2$
  with: $z_1 = x_s^2 + y_s^2 - K_p^2 - A^2 - B^2$; $z_2 = B^2x_s^2 + A^2y_s^2 - A^2K_p^2 - B^2K_p^2 - A^2B^2$ and $z_3 = (ABK_p)^2$.

- (x_s, y_s) correspond to the position of the robot according to the Surrounded Ellipse (SE) center (cf. Figure 2.2).

- $r = 1$ for the clockwise trajectories (cf. Figure 2.4(a)) and $r = -1$ for the counter-clockwise trajectories (cf. Figure 2.4(b)).

- A and B characterize, respectively, major and minor SE axes (cf. Figure 2.2).

- $K_p \in \mathbb{R}^+$ (and $\neq 0$) corresponds to the PELC offset with regard to SE. This offset is equal generally to the dimension of the robot (R_R) (cf. Figure 2.2) plus a certain "Margin" which corresponds to a safety tolerance including perception uncertainty, control reliability and accuracy, etc. Thus, $K_p = R_R + \text{Margin}$.

- $\mu \in \mathbb{R}^+$ a positive constant value which enables us to modulate the convergence of the PELC toward its orbit. The convergence is as slow as μ is smallest (cf. Figure 2.5(a)), which permits us also to obtain smoother PELC.

The PELC given by equation (2.3) could be defined according to a global reference frame (X_GY_G in Figure 2.2). Indeed, it is enough to apply a translation (h, k) and a rotation (Ω) w.r.t. this global reference frame. In Figures 2.4 (a) and (b), the shown PELCs are characterized by: a center $(h, k) = (1, 1)$; an orientation $\Omega = \pi/4$; a half major and minor axes equal respectively to $A = 1$ and $B = 0.25$; an offset $K_p = 0.5$, and finally $\mu = 1$. It is observed that the Parallel Ellipse of Influence (PEI, cf. Figure 2.4) given by $(h, k, \Omega, A, B, K_p) = (1, 1, \pi/4, 1, 0.25, 0.5)$ is a periodic

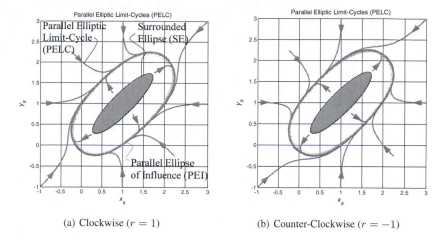

(a) Clockwise ($r = 1$) (b) Counter-Clockwise ($r = -1$)

FIGURE 2.4 Shape possibilities for the used Parallel Elliptic Limit-Cycles.

orbit. The trajectory which has this kind of periodic orbit is called a Parallel Elliptic Limit-Cycle (PELC).

The trajectories from all points of the $X_G Y_G$ global reference frame, including inside the PEI, move toward it (cf. Figure 2.4). To demonstrate analytically the validity of the result for any initial state of the PELC, the following Lyapunov function is used:

$$V(x) = (1/2)(x_s^2 + y_s^2). \tag{2.4}$$

The derivative of $V(x)$ along the trajectories of the system is given by $\dot{V}(x) = x_s \dot{x}_s + y_s \dot{y}_s$. After some development it is obtained finally:

$$\dot{V}(x) = 2\mu V(x)(1 - \Psi). \tag{2.5}$$

The derivative of $\dot{V}(x)$ is:

- negative if $(1 - \Psi) < 0$, thus if the initial condition (x_{s0}, y_{s0}) is outside the PEI (given by equation $\Psi = 1$),

- and positive if the initial condition is inside the PEI $(1 - \Psi) > 0$.

Therefore, the PEI given by $\Psi = 1$ is always the periodic orbit of the PELC. It can be mentioned also from equation 2.5 that, the higher the value of μ is, more the limit-cycle velocity converges toward its periodic orbit (the opposite is true, cf. Figure 2.5(a)).

It is important to notice that the PEI (the PELC orbit attractor, cf. equation 2.3), contrary to a standard Ellipse of Influence (EI, as defined in [Adouane et al., 2011]) permits us to guarantee always an effective minimal distance w.r.t. the contours of the ellipse surrounding the obstacle (SE in Figure 2.5(b)); this constant offset is equal to K_p as given above. Indeed, while using the formulations given in

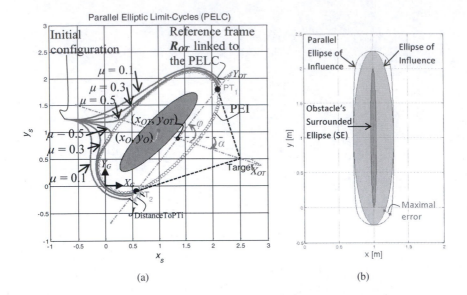

(a)   (b)

FIGURE 2.5   (a) Reference frame linked to each PELC and different PELC shapes according to μ values. (b) Difference between Parallel Ellipse and Ellipse of Influence.

[Adouane et al., 2011], the EI is obtained while fixing its half major A_{lc} and minor B_{lc} axes according to those (A, B) of the SE, it was used $A_{lc} = A + K_p$ and $B_{lc} = B + K_p$. Nevertheless, this formulation permits us to insure accurately this minimal distance (K_p) only in 4 points which correspond to minor and major end-point axes, as given in Figure 2.5(b). Indeed, in this figure it is shown the difference between obtained Parallel EI (PEI) and EI, when: $A = 1m$; $B = 0.05m$ and $K_p = 0.25m$. As maximal difference, an error of $0.1m$ is obtained, corresponding to an error ratio of 40%, which is too high if the aim is to insure an effective safe robot navigation in all situations. Obviously we can increase, according to this maximum error ratio, the value of K_p, but this will not lead to an optimal path, notably in terms of length. It is also important to note that the error ratio is as big as the obstacle is longitudinal (i.e., $A \gg B$).

Therefore, the proposed PELC, in addition to having safest navigation, will permit us to obtain smooth and flexible navigation in different environments (e.g., cluttered or not, structured or not, static or dynamic). For instance, the walls/sidewalks limits of an urban environment could be surrounded by a very thin PEI (cf. section 4.4.4.1.2, page 96). Several simulations and experiments will be shown throughout the different chapters composing this manuscript, for instance, PELC will be used in section 2.5.5 to highlight reactive robot navigation and in section 4.4.4, (page 92) for more cognitive and hybrid (reactive/cognitive) navigation.

Once the mathematical formulation and the interesting features of the proposed PELC are emphasized, let us introduce in the following subsection an appropriate reference frame permitting us to evaluate the pertinence of current achieved sub-tasks.

2.3.2 Reference frames linked to the task achievement

For simple and efficient description of robot navigation in any kind of environment, it is introduced in what follows a specific reference frame assigned for each obstacle / wall / target / etc. inside the considered environment (or at least for each element inside the robot's field of view). These specific references frames will guide the robot behaviors and allow us to evaluate the success of the current achieved sub-task (e.g., wall following, obstacle avoidance, target reaching/tracking, etc.). Each elementary reference frame will orient thus locally the achievement of the robot navigation toward its final objective. A kind of analogy could be established with robot manipulator modeling. In fact, when we would like to control the movement of a robot end-effector (w.r.t. its base), it is assigned for each articulation an appropriate reference frame while using dedicated conventions [Denavit and Hartenberg, 1955] [Khalil and Dombre, 2004]. These local reference frames are mainly used to express simply the local elementary articulations' movements (translation / rotations) in order to obtain the desired final end-effector movement. The context of robot navigation is obviously different but the proposed reference frames will help to make a reasoning on the efficiency of the robot movements in order to reach its final objective. To define such specific reference frames it is mandatory to fix the center and the orientation of its axes.

Let us start to define the reference frames linked to each possible obstacle (wall / pedestrian or any object which could obstruct the robot's movement). They will have a specific reference frame ($\mathfrak{R}_{OT}$) permitting us to define the obstacle avoidance sub-task achievement (or set-points) while knowing the localization of the robot according to it. $\mathfrak{R}_{OT}$ is obtained with a simple geometric construction and has the following features (cf. Figure 2.5(a)):

- X_{OT} axis connects the center of the obstacle (x_O, y_O) to the center of the final Target (x_f, y_f). This axis is oriented toward this target.

- Y_{OT} axis is defined by two points PT_1 and PT_2, which correspond to the tangent points between the two straight lines coming from the final target (x_f, y_f) and the PEI. Y_{OT} axis is oriented while following trigonometric convention.

The axes X_{OT} and Y_{OT} intersect on the point $O_{OT} = (x_{OT}, y_{OT})$ and they have an angle φ between them. To guide the robot's future movements, it is important to define its localization w.r.t. $\mathfrak{R}_{OT}$. One needs, therefore, to make a transformation from the global reference frame $X_G Y_G$ to the local reference frame $X_{OT} Y_{OT}$. Knowing that $X_{OT} Y_{OT}$ is not necessarily orthogonal, it proceeds with two steps:

1. Transformation of the robot localization $(x, y)_G$ from the global reference frame to an intermediate orthonormal reference frame which has as center (x_{OT}, y_{OT}) and X axis $= X_{OT}$. The following homogeneous transformation

is used (cf. Figure 2.5(a)):

$$
\begin{pmatrix} x \\ y \\ 0 \\ 1 \end{pmatrix}_{OT_I} = \begin{bmatrix} \cos\alpha & -\sin\alpha & 0 & x_{OT} \\ \sin\alpha & \cos\alpha & 0 & y_{OT} \\ 0 & 0 & 1 & 0 \\ 0 & 0 & 0 & 1 \end{bmatrix}^{-1} \begin{pmatrix} x \\ y \\ 0 \\ 1 \end{pmatrix}_G \tag{2.6}
$$

2. Once x_{OT_I} and y_{OT_I} are obtained for this intermediate transformation, one can obtain finally the localization of the robot (x_{RO}, y_{RO}) w.r.t. $\mathfrak{R}_{OT}$ as follows (cf. Figure 2.5(a)):

$$
\begin{cases} x_{RO} = x_{OT_I} - y_{OT_I} \dfrac{\cos(\varphi - \alpha)}{\sin(\varphi - \alpha)} \\[2mm] y_{RO} = \dfrac{y_{OT_I}}{\sin(\varphi - \alpha)}. \end{cases} \tag{2.7}
$$

It is to be mentioned that $(\varphi - \alpha) \neq 0$ modulo π. It is always true because the axes X_{OT} and Y_{OT} are never parallel (or co-linear) from the moment that (x_f, y_f) is outside of the PEI attributed to the considered obstacle and the obstacle exits effectively (A and B features $\neq 0$). According to that, $sin(\varphi - \alpha)$ (used in Equation 2.7) is always $\neq 0$.

The main idea to use this essential component ($\mathfrak{R}_{OT}$) is to determine which PELC the robot must follow to avoid for example an obstacle. In fact, once the transformation from the global frame $X_G Y_G$ to the local reference frame $X_{OT} Y_{OT}$ is done, it is enough for instance to check the sign of the robot's localization w.r.t. the axis X_{OT} to assign the appropriate robot behavior [Adouane et al., 2011]. For instance, if the sign of x_{RO} is negative, the robot must follow the defined PELC (to avoid the obstacle) and if positive the robot can consider that the obstacle is not an obstructing obstacle and can go therefore straight toward to its final target (cf. Figure 2.5(a)). At the condition obviously that there is no other constrained obstacle; if not, the process will be reiterated. In section 4.3 (page 80), the optimality of the followed PELC will be presented while determining its best direction and shape. This will be done notably while obtaining the optimal value of μ (cf. equation 2.3).

Furthermore, the final target $T = (x_f, y_f)$ (which could not have any particular orientation in the environment) will be assigned also a reference frame ($\mathfrak{R}_T$) which has as center (x_f, y_f) and as axes, orthogonal one, oriented just as the global reference frame axes. This description permits the homogenization of using these references frames. It is to be noted that this kind of reference frame $\mathfrak{R}_{T_i}$ will be notably used in the planning method presented in section 4.3 (page 80) and in chapter 5 where each intermediate target in the environment will be attributed an appropriate reference frame to sequentially guide the navigation of the robot (from one intermediate target to another).

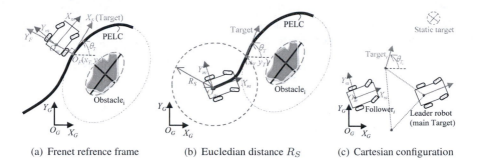

(a) Frenet refrence frame   (b) Eucledian distance R_S (c) Cartesian configuration

FIGURE 2.6 Set-points definition based on (a) global planned path, (b) local planned path, (c) general static/dynamic target.

2.4 HOMOGENEOUS SET-POINTS DEFINITION FOR ROBOT'S NAVIGATION SUB-TASKS

The design of multi-controller architecture aims to decompose the overall complex task into a multitude of sub-tasks to achieve (e.g., target reaching, wall following, obstacle avoidance, etc.) (cf. section 1.4, page 15). According to these elementary sub-tasks, performed in a reactive or cognitive way (cf. section 1.3.2), it has been noted in general that the robot must follow/track a path/trajectory or reach/track a specific target. This section aims to propose a homogeneous set-points definition for the multitude of the robot's navigation sub-tasks in order to simplify the design of control architectures.

As described in section 1.5 (page 21) the use of static or dynamic targets could lead to a much more flexible way to define the robot's sub-tasks, it is promoted therefore in our works the use of target set-points, defined by a pose (x_T, y_T, θ_T) and a velocity v_T. The following subsections (2.4.1 to 2.4.3) will highlight the fact that this set-point formulation is generic enough to define an important number of the robot's behaviors. It is to be noted that, once the set-points are defined, at each sample time, it is important to have reliable control laws to reach/track these assigned set-points. To do that, one of the reliable control laws defined in section 3.2 (page 54) will be used to stabilize the errors to zero.

2.4.1 Target tracking set-points based on global planned path

The first identified case corresponds to the one where a global path is already defined using for instance PELC (cf. section 4.3, page 80). In fact, in certain situations (e.g., static environment) it is enough for the robot to follow the path as accurately as possible without modifying its initial planning. In that situation, a Frenet reference frame is used [Samson, 1995] to extract the robot's set-points. The target set-point, at each sample time, is given by (cf. Figure 2.6(a)):

- A position (x_T, y_T) corresponding to the closest position in the pre-planned path w.r.t. the origin of the reference frame $X_m Y_m$. (x_T, y_T) point corresponds to the origin of Frenet reference frame $X_F Y_F$.

- An orientation θ_T corresponding to the tangent of the path w.r.t. $X_G Y_G$ reference frame.

- A velocity v_T which could be constant or variable indifferently.

2.4.2 Target tracking set-points based on local planned path

Here, the set-point configurations are taken within the generated PELC trajectories (cf. subsection 2.3.1) but the same principle could be used for any other online local generated trajectory obtained from local planners.

When the environment is not very well known or dynamic, it is better to navigate reactively (cf. section 1.3.2, page 11). In that situation, the current PELC takes as initial configuration, and at each sample time, the current robot configuration. The target set-point is given by (cf. Figure 2.6(b)):

- A position (x_T, y_T) corresponding to the intersection between the circle (which has as origin the origin of the reference frame $X_m Y_m$ and as radius R_S) and the planned PELC.

- An orientation θ_T corresponding to the tangent to the PELC w.r.t. $X_G Y_G$ reference frame at the intersection point (x_T, y_T). If $R_S = 0$, the robot has to apply only an orientation control. Indeed, since the robot is already on the current computed PELC, the robot has only to control its heading w.r.t. θ_T. This simple control has been used in [Adouane, 2009b] and [Adouane et al., 2011].

- A velocity v_T which could be constant or variable indifferently.

2.4.3 General target reaching/tracking set-points

The last identified case (cf. Figure 2.6(c)) corresponds to the general situation where the robot must reach/track a static/dynamic target $(x_T, y_T, \theta_T, v_T)$. The sub-tasks which can deal with this kind of target definition correspond to all the cases where the set-points are not restricted to evolve inside a specific path. For instance, let us cite:

- For a static target, the set-points could correspond to the final robot destination as given in Figure 2.2. They could also correspond to an appropriate waypoints in the environment through which the robot must cross sequentially (chapter 5 will give a complete example showing this navigation strategy).

- For a dynamic target, this kind of target set-point can serve for the Follower robot (as depicted in Figure 2.6(c)) to track a secondary target referenced w.r.t. the Leader. Chapter 6, focused on multi-robot systems, will highlight better this kind of target set-point definition.

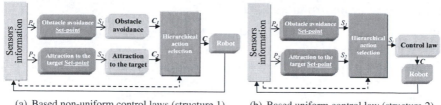

(a) Based non-uniform control laws (structure 1)   (b) Based uniform control law (structure 2)

FIGURE 2.7   (See color insert) The two main used multi-controller structures.

2.5   MULTI-CONTROLLER ARCHITECTURES FOR FULLY REACTIVE NAVIGATION

2.5.1   Main structures

It will be presented, in what follows, two simple multi-controller architectures (cf. Figure 2.7) which will serve as basic structures to highlight the main components to perform flexible and reliable navigation in cluttered environment. While exploiting the principle of bottom-up construction characterizing these control architectures (cf. section 1.4, page 15), the sophistication of these simple architectures will be increased, permitting us, for instance, in chapter 3 to analyze the overall stability and smoothness of the robot's navigation and in chapter 4 to add high-level features (planning and re-planning).

Before highlighting the main difference between the two control structures depicted in Figure 2.7, it is important to give the definition of the robot's elementary behavior (controller). In what follows each robot's controller is constituted by a dedicated set-point and stable control law blocks which permit us to achieve safely and reliably the desired robot's behavior. As shown in Figures 2.7(a) and 2.7(b) the two structures 1 and 2, are different in terms of the used control laws. In fact, structure 1 has two distinct control laws whereas structure 2 has only one common control law shared by the two set-point blocks.

It is interesting to notice that structure 1 is the less restrictive architecture, in the sense that both controllers (behaviors) can have completely different set-points and control laws definition. It is enough to have two already stable elementary controllers to integrate them in this structure, without any harmonization of the used set-points or control laws. The possible drawback of this kind of multi-controller architecture corresponds to its difficulty in having a simple analysis of the overall control architecture stability (since it could use non-uniform control laws). This aspect will be widely discussed in chapter 3.

The two simple multi-controller architectures depicted in Figure 2.7 permit us to manage the interaction between different elementary blocks. The main features of each block composing these architectures are detailed below.

2.5.2 Sensor information block

While using robot's sensors and any already known data on the environment (using a road-map for instance), this block is in charge of detecting / localizing / characterizing any important features in the environment. Mainly this block in the case of the basic architectures given in Figure 2.7 must provide the list of all perceived obstacles and the relative final target localization w.r.t. the robot (cf. Figure 2.2).

Any possible obstructing object (obstacle / wall / pedestrian / etc.) is characterized as specified in section 2.1 by a Surrounded Ellipse (SE) given by the parameters $(h, k, A, B$ and $\Omega)$ (cf. equation 2.1 and Figure 2.2). Even if the perceptive aspect is not the main topic addressed in our research works, nevertheless knowing that this characterization plays an important role to perform reliable safe navigation, some works have been done. Different techniques have been proposed in the literature to enclose uncertain data with an ellipse [Porrill, 1990] [Welzl, 1991] [Zhang, 1997] [De Maesschalck et al., 2000]. In [Vilca et al., 2012b], a review of different methods of enclosing an ellipse is given. This characterization can be insured even offline (using for instance a road map) or online using for example a camera positioned in the environment [Benzerrouk et al., 2009] or the robot's infrared sensors [Vilca et al., 2012c]. Obviously, among the most challenging aspects linked to obstacle detection and characterization is the case where the robot has to discover the environment online and with only its own sensors [Vilca et al., 2013a] (it is the case for fully reactive navigation) or while merging locally the data given by several robots [Lozenguez et al., 2011a] [Vilca et al., 2012a]. The data segmentation remain also an important problem to address [Rodriguez, 2014].

In [Vilca et al., 2012b] SE parameters have been obtained online as the robot moves (cf. Figure 2.8). In this work, several methods, on the acquired sequence of

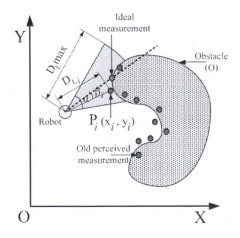

FIGURE 2.8 N Range data set expressed using polar coordinate $(D_{Li}, \beta_i) \mid_{i=1..N}$ w.r.t. the robot.

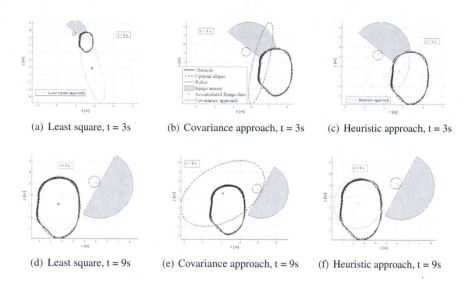

(a) Least square, t = 3s   (b) Covariance approach, t = 3s   (c) Heuristic approach, t = 3s

(d) Least square, t = 9s   (e) Covariance approach, t = 9s   (f) Heuristic approach, t = 9s

FIGURE 2.9 Evolution of the obtained Surrounded Ellipses (SE) using the different studied approaches.

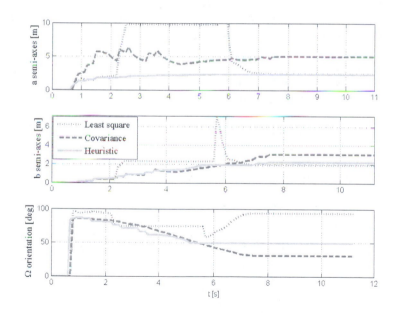

FIGURE 2.10 Evolution of the estimated parameters of the surrounded ellipse.

uncertain telemetric range data, have been used. Figures 2.9 ((a), (d)); ((b), (e)) and ((c), (f)) show the obtained SE at two instants (t=3s and 9s) using, respectively, least square, covariance and heuristic methods. Furthermore, it is observed in Figure 2.10 that SE parameters of the least square and covariance methods can change abruptly whereas the obtained ellipse parameters, with heuristic method, change much more smoothly while permitting us to better enclose the detected obstacle (cf. Figure 2.9). The smoothness of the evolution of the identified ellipse's parameters permits at its turn to perform smooth robot avoidance in a cluttered environment. The extension of this work has been done in [Vilca et al., 2012c] [Vilca et al., 2013a] using the Extended Kalman Filter (EKF) and an appropriate sub-optimal heuristic method to insure that all the acquired uncertain range data are actually filtered and surrounded by the computed ellipse and that the evolution of SE parameters remains always smooth.

Section 2.5.5 will show some results on the use of these perceptive techniques as inputs for a reactive obstacle avoidance controller. This perceptive block will be used also for instance in the different following chapters as in section 5.5.2 (page 135) where a group of real VipaLab vehicles have to avoid an obstacle.

2.5.3 Hierarchical action selection block

The coordination of the several controllers constituting a multi-controller architecture is among the most important aspect to master in order to obtain coherent and reliable architecture (cf. section 1.4, page 15). Chapter 3 will give more details about the different developed techniques to ensure the overall reliability and smoothness of such architectures.

In the simple architectures depicted in Figure 2.7, the activation of one controller in favor of another is achieved completely in a hierarchical way like the principle of subsumption proposed initially by Brooks in [Brooks, 1986]. Algorithm 1 gives the logic of controllers' activation ("Obstacle avoidance" and "Attraction to the target"). In summary, the "Hierarchical action selection" block activates the obstacle avoidance controller as soon as there exists at least one obstacle which can obstruct the robot's future movements toward the final target. Therefore, the robot's local stimuli are only responsible to trigger the switch between controllers.

Algorithm 1: Hierarchical action selection

Input: All the features (h, k, Ω, A, B) of the closest constrained obstacle (cf. Algorithm 2); Value of K_p (the desired minimum safe distance "offset" to the obstacles); Current final target localization (x_f, y_f).

Output: Define the controller to activate

1 **if** It exists at least one obstructing obstacle (cf. Algorithm 2) **then**

2   Activate *Obstacle avoidance* controller (cf. Algorithm 3)

3 **else**

4   Activate *Attraction to the target* controller

5 **end**

Algorithm 2: Obtaining the most obstructing obstacle

Input: All the features $(h, k, \Omega, A, B)_i$ (cf. equation 2.1) of the obstacles in the robot's field of view; Value of K_p (the desired minimum safe distance "offset" to the obstacles); Current final target localization (x_f, y_f).
Output: The index "k"of the most obstructing obstacle (if it exists).

1 **for** Each Obstacle$_i$ **do**
2   **if** The obstacle is an obstructing obstacle
     {i.e., it exists one intersect point between the line "l" (connecting the robot to the target (cf. Figure 2.2)) and the Parallel Ellipse of Influence of the Obstacle$_i$ **then**
3 | Add the Obstacle$_i$ to $ListObstructingObstacles$;
4   **end**
5 **end**
6 **if** $ListObstructingObstacles \neq \emptyset$ **then**
7   Extract the index $k \in ListObstructingObstacles$ of the closest Euclidean distance to the robot D_{ROi} (cf. Figure 2.2).
8   **if** Two or more obstructing obstacles have the same value of D_{ROi} **then**
9   choose to avoid the one with the smallest D_{PROi} (cf. Figure 2.2)
10   **if** It is already the same D_{PROi} **then**
11 | choose arbitrary one of these obstacles.
12   **end**
13   **end**
14 **else**
15   There is not obstructing obstacle, the environment is safe.
16 **end**

2.5.4   Set-point blocks

These blocks, which have as input the perceptions P_i coming from the sensor information block (cf. subsection 2.5.2), are responsible to give for each dedicated controller, the appropriate set-points for its operation.

As given in section 2.4, the adopted homogeneous set-point definition for each controller (sub-task) has the following data $(x_T, y_T, \theta_T, v_T)$, which gather the pose and the velocity of the desired target to reach/track. In what follows, the set-points definition, dedicated to the "Obstacle avoidance" and "Attraction to the target" controllers, will be given.

2.5.4.1   Reactive obstacle avoidance controller

The objective of this controller is to avoid with high reactivity (with the less computation time and without high-level cognition (cf. section 1.3.2, page 11)) any obstacle hindering the robot's movement toward its final target. In what follows only a simple formulation of PELC (cf. section 2.3.1) will be used to perform reactive obstacle avoidance. The actual potentialities of the proposed PELC will be illustrated notably in chapter 4. Indeed, in section 4.3 (page 80) optimal PELC* is used to perform sophisticated navigation in a cluttered environment while using either local or global planning. In the current chapter, the PELC are used to perform fully reactive navigation. In fact, contrary to PELC*, the reactive navigation given in this section will use PELC with fixed value of μ (cf. equation 2.3) (and not with optimal μ^* as for PELC*). Besides, the used reference frame (cf. section 2.3.2) and the obstacle avoidance behavior will use only simple rules. Despite all the applied simplifications, subsection 2.5.5 will show several simulations and experiments emphasizing the efficiency of the proposed fully reactive navigation in a cluttered environment.

The reactive obstacle avoidance defined in Algorithm 3 is developed according to stimuli-response principle. To implement this kind of fully reactive obstacle avoidance behavior it is important at least to:

- detect the obstacle to avoid (cf. Algorithm 2),

- decide the direction of the avoidance (clockwise or counter-clockwise), and

- define an escape criterion which defines if the obstacle is completely avoided or not yet.

All these different elements must be defined and applied while guaranteeing that the robot trajectory is safe, smooth and avoids undesirable situations as deadlocks or local minima; and that the stability of the applied control law is guaranteed. The necessary steps to carry out this fully reactive obstacle avoidance algorithm are given below:

1. For each sample time, run Algorithm 2 to obtain the index "k" of the most obstructing obstacle.

2. After the determination of the closest obstructing obstacle, we need to determine the robot reflex behavior w.r.t. this obstacle: clockwise or counter-clockwise avoidance; repulsive or attractive phase (cf. Algorithm 3). The four possible robot behaviors corresponding to 4 specific robot localizations (areas as given in Figure 2.11) are distinguished while using an orthogonal reference frame. As used in [Adouane, 2008] or [Adouane et al., 2011], the reference frame $\mathfrak{R}_{OT}$, linked to each obstacle, has been proposed so that its determination becomes the most simple, permitting therefore an even more reactive navigation. This specific reference frame has the following features (cf. Figure 2.11):

   - As given in the general case (cf. Figure 2.5(a)), X_{OT} axis connects the center of the obstacle (x_O, y_O) to the center of the final Target (x_f, y_f). This axis is oriented toward this target.

   - Y_O axis is perpendicular to the X_O axis and is oriented while following a trigonometric convention.

3. Since the obtained reference frame $\mathfrak{R}_{OT}$ is always orthonormal, to obtain the robot localization w.r.t. $\mathfrak{R}_{OT}$ it is enough to apply the first transformation given in equation 2.6. According to this localization, the robot can deduce more readily its reactive behavior. For instance, the sign y_{RO} (ordinate of the

robot in $\mathfrak{R}_{OT}$) has been used to determine the suitable direction of obstacle avoidance. If $y_{RO} \geq 0$, then apply clockwise limit-cycle direction, otherwise apply counter-clockwise direction. Nevertheless, this direction is forced to the direction taken just before if the obstacle avoidance controller was already active at $(t - \delta T)$ instant and this is to avoid local minima and dead-ends [Adouane, 2009b]. These simple rules permit us to reduce the length of robot trajectory to reach its final target. Furthermore, it is seen in Algorithm 3 that the sign of x_O is used also to determine if the robot is in the repulsive or at-

---

Algorithm 3: Reactive set-point definition for reactive obstacle avoidance based PELC

Input: All the features of the closest constrained obstacle$_i \equiv (h, k, \Omega, A, B)_i$; Values of μ and the desired offset K_p' (cf. equation 2.3); Current final target localization (x_f, y_f).

Output: Set-point $(x_T, y_T, \theta_T, v_T)$ to reach based on the PELC to follow.

// **I) Obtaining the PELC offset "K_p" of the PELC to follow**

1 **if** $x_O \leq 0$ **then**

2      $K_p = K_p' - \xi$ **(Attractive phase)**

3   {with ξ a small constant value as $\xi \ll Margin$ (cf. subsection 2.3.1) which guarantees that the robot does not navigate very closely to the desired Parallel Ellipse of Influence (PEI) (which could causes the oscillations of the robot's trajectory [Adouane, 2009b])}

4 **else**

5   {Escape criterion: go out of the obstacle's PEI with smooth way}

6      $K_p = K_p + \xi$ **(Repulsive phase)**

7 **end**

// **II) Obtaining the PELC direction (clockwise or not, "r" in equation 2.3)**

8 **if** obstacle avoidance controller was active at $(t - \delta T)$ instant **then**

9   Apply the same direction already used before $r = r_{previous}$.

10   {This will permit to avoid some conflicting situations leading to robot dead-ends [Adouane, 2009b]}

11 **else**

12      $r = sign(y_{RO})$ (cf. equation 2.6)

13 **end**

// **III) Set-point for reactive obstacle avoidance**

14 {Knowing the parameters of the closest constrained obstacle$_i$, the value of μ and according to the obtained K_p and r values, the PELC to follow (thus the evolution $(\dot{x}_s, \dot{y}_s)$) is completely defined using equation 2.3. While taking the value of $R_S = 0$ in subsection 2.4.2 (cf. Figure 2.6(b)), the target to track has the following features:}

$$\begin{cases} (x_T, y_T) = (x, y) \text{ //The current robot location} \\ \theta_T = \arctan(\frac{\dot{y}_s}{\dot{x}_s}) \\ v_T = \text{constant //} v_T \text{ is chosen constant for simplification} \end{cases}$$

15

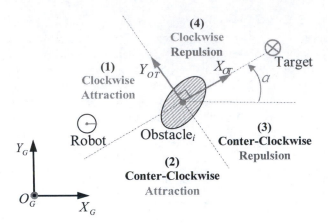

FIGURE 2.11 The four specific areas surrounding the obstacle to avoid.

tractive phase (cf. Figure 2.11). In repulsive phase, the limit-cycle takes an increased value of K_p to guarantee the robot's trajectory smoothness. Several examples are given in subsection 2.5.5 to show the efficiency of the presented fully reactive navigation in a cluttered environment.

In addition, it is important to mention that for an even more simplicity of the use of Algorithm 3, the PELC can be described with its simplest form corresponding to a circular limit-cycle. In fact, in equation 2.3, if $A = B$ (which means that the detected obstacle is surrounded with a circular shape), the PELC can be simplified and will be given by the following equation [Adouane, 2009b]

$$\dot{x}_s = ry_s + \mu x_s(R_c^2 - x_s^2 - y_s^2)$$
$$\dot{y}_s = -rx_s + \mu y_s(R_c^2 - x_s^2 - y_s^2)$$

(2.8)

where $R_c = R_O + K_p$ is the radius of the circle of convergence and R_O the radius of the circle surrounding the detected obstacle. Figure 2.12 shows the obtained limit-cycle when $R_c = 1$. The trajectories from all points (x_s, y_s) including inside the circle, move toward the circle. It is to be mentioned that to surround the detected obstacle by a circle shape, either $R_O = A = B$ (the obstacle is already circular) or R_O is set to A (the major axis value of SE (cf. Figure 2.2)).

As in any reactive navigation (cf. Algorithm 3), it is important to manage, with simple rules, some conflicting situations due principally to the fact that reactive navigation supposes local reaction to the environment stimuli and not according to global or sophisticated environment knowledge nor high cognition level (cf. section 1.3.2, page 11). The specific local and reactive rules to avoid local minima and/or dead ends are detailed in [Adouane, 2009b].

Once the reactive obstacle avoidance set-points are obtained (cf. Algorithm 3) it is enough to use one of the stable control laws defined in section 3.2 (page 54), for target tracking. The choice of the control law depends obviously on the structure of

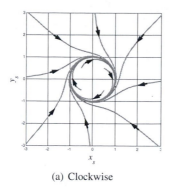

(a) Clockwise

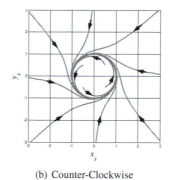

(b) Counter-Clockwise

FIGURE 2.12 Different shapes for circular limit-cycles.

the used robot (unicycle (cf. subsection 3.2.1.2) or tricycle (cf. subsection 3.2.2)). In addition, according to obtained set-points in Algorithm 3, the used control therefore is simplified to be an orientation control [Adouane et al., 2011].

2.5.4.2   Attraction to the target controller

To obtain the set-points dedicated to this controller, it is enough to know the final target position (x_f, y_f); the set-point is given therefore by: $(x_T, y_T, \theta_T, v_T) = (x_f, y_f, 0, 0)$. In fact, the aim of this controller is to reach a static target ($v_T = 0$) with any value of angle, θ_T is then chosen $= 0$.

Once the set-point is obtained, it is enough here also to use one of the stable control laws defined in subsection 3.2 (page 54) for target reaching. It is important to notice that according to the definition given above for "Attraction to the target" and "Obstacle avoidance" controllers, either structure 1 or 2 (cf. Figure 2.7) could be used to perform the targeted navigation. Stability analysis of both multi-controller structures, while taking into account the transition phase between controllers, will be detailed in chapter 3.

In what follows, several simulations and experiments will show the relevance of the proposed multi-controller architectures for reactive navigation in a cluttered environment.

2.5.5   Simulations and experimental results

To demonstrate the efficiency of the proposed multi-controller architecture (cf. section 2.5.1) several simulations and experiments have been made. Some of them are given below.

The first simulation will emphasize notably the performance of the proposed reactive obstacle avoidance method-based limit-cycles (cf. subsection 2.5.4.1) and the online obstacle detection and characterization method (cf. subsection 2.5.2). A mobile robot with a radius of $R_R = 0.065\ m$ and six infrared range sensors, with the

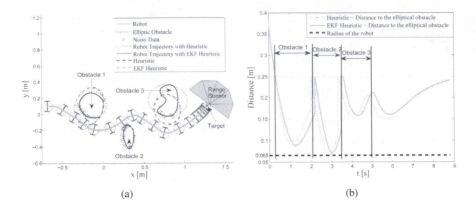

(a)   (b)

FIGURE 2.13 (a) Robot trajectory using fully reactive navigation and the proposed heuristic and EKF approach to enclosing the obstacles [Vilca et al., 2013a]. (b) Distance between the robot and elliptical obstacles.

maximum detected range equal to $D_L max = 0.30$ m is considered (cf. Figure 2.8). These sensors are in the front of the robot, with $30°$ between each pairs of sensors. The accuracy of the used sensors based on the data-sheet is around 10% of $D_L max$. It is considered in this simulation an uncertainty range with a maximum value of 20% of $D_L max$, ensuring thus to take the worst range value.

Figure 2.13(a) shows the robot trajectory in an environment with three obstacles (clockwise and counter-clockwise avoidance are observed). The red points represent the range data given by the sensors along all the robot movement. The range data buffer used to compute the ellipse parameters are deleted for each new discovered obstacle. Figure 2.13(b) represents the minimum distance between the effective elliptical obstacles[2] and the robot position along its trajectory using either the heuristic method (red dotted line) or the combination of heuristic method and EKF (green continuous line). This last figure confirms the non-collision of the robot with any obstacle and better safety when the combination of heuristic method and EKF are used. In general, the simulation depicted in Figure 2.13 confirms the safety and the smoothness of the robot navigation even if the obstacles are discovered and characterized online [Vilca et al., 2013a].

Furthermore, to show the efficiency of the proposed fully reactive navigation, a statistical survey has been made while doing a large number of simulations in different cluttered environments. We did specifically 1000 simulations where each uses several dozen obstacles with different random positions in the environment (cf. Figure 2.14 for two examples of performed simulations). It is to be noted that each obstacle is subject to parameter uncertainty, representing the inaccuracy of robot's infrared sensors. 97% of the performed simulations permitted the robot to reach the target in a smooth way and in finite time, while avoiding collisions, local minima

---

[2]Obtained while knowing all the range data, without noise, surrounding the obstacle.

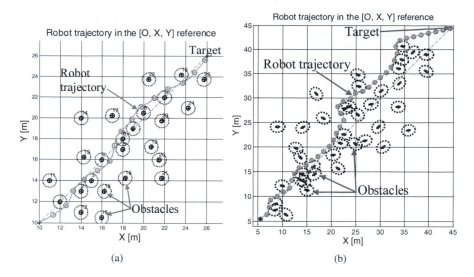

FIGURE 2.14 Smooth robot trajectories obtained with the proposed reactive navigation while using (a) circular limit-cycles (b) elliptic limit-cycles.

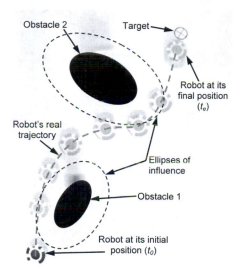

FIGURE 2.15 Top view of the robot trajectory using the developed platform (cf. Annex A).

and dead-ends (cf. Figure 2.14). This is an encouraging result compared to the re-strictions/constraints imposed on reactive navigation such as no planning step, no global knowledge about the environment, etc. Three percent of the failed simulations are due mainly to some specific obstacle configurations (no free path solution between the robot and the target) and to the large amount of introduced noise.

In addition, several experiments using Khepera® robots (cf. Annex A) have been made. Figure 2.15 shows the first experiments where the robot's localization and obstacle features were obtained thanks to a camera positioned in the top of the experimental platform (cf. Annex A). It can be seen that the robot successfully converges to its target at moment t_e after avoiding two obstacles surrounded with two ellipses of influence [Adouane et al., 2011]. Figure 2.16 shows another experiment where the robot uses its own infrared sensors and the proposed method-based EKF to characterize the detected obstacles [Vilca et al., 2013a]. This last experiment highlights the effectiveness of the proposed fully reactive navigation in a cluttered environment.

Other experiments performing navigation in a cluttered environment, using Khepera or VipaLab robots, will be shown notably in chapters 5 and 6.

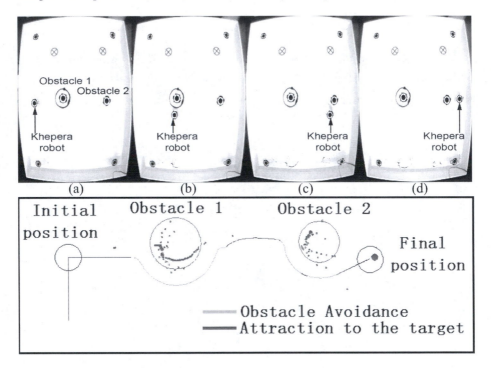

FIGURE 2.16 Top view of the robot trajectory and the observed infrared uncertain range data.

2.6 CONCLUSION

This chapter introduced the general framework of the targeted autonomous robot navigation. It emphasized two main controllers ("obstacle avoidance" and "Attraction to the target") which will be widely used throughout this manuscript.

This chapter also proposed a homogeneous set-points definition based on target reaching/tracking $(x_T, y_T, \theta_T, v_T)$ while highlighting the fact that these set-points are generic and flexible enough to define an important number of the robot's sub-tasks. The aim of set-point harmonization is, among other things, to simplify the design of multi-controller architectures promoted in this manuscript. Indeed, having generic set-point, permits us to have dedicated control laws to deal with, and to avoid having in the same control architecture, several set-point conventions and therefore also several control laws. This last case could lead to complex stability analysis. The stability of each controller as well as the overall multi-controller architecture will be addressed in the following chapter.

A large part of this chapter has been dedicated to defining the proposed obstacle avoidance controller, which is an important component to perform safe robot navigation in a cluttered / dynamic environment. The main elements of this controller are:

- Safe orbital trajectories, called PELC (Parallel Elliptic Limit-Cycle), permitting navigation near obstacles while letting them always be at a minimal defined distance. The mathematical formulation of the PELC was presented and motivated in this chapter.

- Specific reference frames assigned to each obstacle / target / etc. inside the considered environment. They permit us to guide the robot's behaviors and to evaluate the success of the current achieved sub-task.

These two elements will be intensively used through all the manuscript to perform either reactive or cognitive navigation in different constrained environments.

Furthermore, a complete multi-controller architecture, dedicated to fully reactive navigation in cluttered environments has been presented. A brief description of the online methodology to detect and to characterize obstacles has been also given. Several simulations and experiments were given to demonstrate the effectiveness of the proposal.

CHAPTER 3

Hybrid$_{CD}$ (continuous/discrete) multi-controller architectures

CONTENTS

THIS CHAPTER focuses on the proposed Hybrid$_{CD}$ (continuous/discrete) multi-controller architectures for online mobile robot navigation in cluttered environments. The developed stable control laws for target reaching/tracking are presented. An important part of this chapter emphasizes how to obtain stable and smooth switching between the different elementary controllers composing the proposed architectures.

3.1 INTRODUCTION

The main investigated issue in this chapter relies on the mean to attest if the realization of an overall complex task (such as autonomous navigation in cluttered and dynamic environment) is globally efficient (e.g., safe, smooth, reliable, etc.). Indeed, it is relatively simple to confirm the reliability, for instance, of an elementary obstacle avoidance, but if the objective is also to maintain a formation in a group of robots and so on, the global analysis quickly becomes complex. This is due mainly to antagonist sub-tasks to achieve or to heterogeneous control variables/set-points. The raised automatic control question could be formulated as "*Is it possible to formalize an analytic function (control law) to guarantee the efficiency of the overall performed complex task?*" This function remains highly difficult to obtain and it is still a challenging issue to define such function in the context of autonomous mobile robotics [Benzerrouk, 2010, chapter 3].

As shown and motivated in section 1.4 (page 15), the undertaken methodology to break the complexity of the different targeted tasks is to proceed with a bottom-up approach. This is done in the presented work through the developments of multi-controller architectures. The potentialities of Hybrid$_{CD}$ systems [Branicky, 1998] [Zefran and Burdick, 1998] [Liberzon, 2003] are taken as a formal framework to demonstrate the overall stability and smoothness of such architectures (cf. section 1.4.2, page 19).

Before giving more details about the smoothness and the stability of the overall proposed multi-controller architectures (cf. section 3.3), let us give in the following section some details about the stable controllers constituting these architectures.

3.2 ELEMENTARY STABLE CONTROLLERS FOR TARGETS REACHING/TRACKING

In the different investigated works, two main robotics structures have been used (unicycle and tricycle). These two structures are among the most common for mobile robotics applications. It is natural therefore to define dedicated stable control laws for these two structures. It is to be noted that stability, in the sense of Lyapunov [Khalil, 2002] (cf. Annex B, page 191), has been used to synthesize the different proposed control laws.

After reviewing the effective model corresponding to each structure, the proposed control laws will be summarized [Adouane, 2009a] [Benzerrouk et al., 2014] [Vilca et al., 2015a].

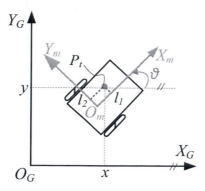

FIGURE 3.1 Generic kinematic model of a unicycle according to a global Cartesian reference frame $X_G Y_G$.

3.2.1 Dedicated controllers to unicycle mobile robots

The generic kinematics model of a unicycle, controlled according to a point "P_t", not specifically positioned between the robot wheels (cf. Figure 3.1), is given by:

$$\dot{\xi} = \begin{pmatrix} \dot{x} \\ \dot{y} \\ \dot{\theta} \end{pmatrix} = \begin{pmatrix} \cos\theta & -l_2\cos\theta - l_1\sin\theta \\ \sin\theta & -l_2\sin\theta + l_1\cos\theta \\ 0 & 1 \end{pmatrix} \begin{pmatrix} v \\ w \end{pmatrix} \tag{3.1}$$

where:

- x, y, θ: configuration state of the unicycle at the point "P_t" of abscissa and ordinate (l_1, l_2) according to the mobile reference frame $X_m Y_m$.

- v and w correspond respectively to the linear and angular velocities of the robot at the point "P_t."

Two main control laws are used for unicycle structure to reach/track targets in the environment. These control laws will be used in next sections and chapters to show the different proprieties of the proposed control architectures.

3.2.1.1 Simple control law for static target reaching

The presented control law will be used exclusively in section 3.3.1 to emphasize the behavior of one of the proposed hybrid$_{CD}$ control architectures. This control law (cf. equation 3.3) permits us to guide the robot toward a final static target $T \equiv (x_T, y_T, \theta_T = 0, v_T = 0)$ while controlling the robot position at the point $P_t = (l_1, 0)$ (cf. Figure 3.1). If this target corresponds to a final target as given in Figure 2.2 (page 27), therfore $(x_T, y_T) = (x_f, y_f)$ and knowing that the target is considered circular with R_T radius, l_1 must be $\leq R_T$ to guarantee that the robot center reaches the final target with asymptotic convergence [Adouane, 2009a]. The robot's kinematics configurations are given by:

$$\begin{pmatrix} \dot{x} \\ \dot{y} \end{pmatrix} = \begin{pmatrix} \cos\theta & -l_1\sin\theta \\ \sin\theta & l_1\cos\theta \end{pmatrix} \begin{pmatrix} v \\ w \end{pmatrix} = M \begin{pmatrix} v \\ w \end{pmatrix} \tag{3.2}$$

with M invertible matrix.

The errors of position are given by: $e_x = x - x_T$ and $e_y = y - y_T$. Knowing that the target is invariable according to the global reference frame, thus: $\dot{e}_x = \dot{x}$ and $\dot{e}_y = \dot{y}$. Classical techniques of linear system stabilization can be used to asymptotically stabilize the error to zero [Laumond, 2001]. A simple proportional controller is used given by:

$$\begin{pmatrix} v \\ w \end{pmatrix} = -KM^{-1}e = -K \begin{pmatrix} \cos\theta & \sin\theta \\ -\sin\theta/l_1 & \cos\theta/l_1 \end{pmatrix} \begin{pmatrix} e_x \\ e_y \end{pmatrix} \tag{3.3}$$

with $K > 0$ and $l_1 \neq 0$.

To demonstrate the stability of the control law given by equation 3.3, it is enough to define the following Lyapunov function V, such as:

$$V = \tfrac{1}{2}d^2 \tag{3.4}$$

with $d = \sqrt{e_x^2 + e_y^2}$ (distance robot-target). Thus, to guarantee the asymptotic stability of this control law, $\dot{V}$ must be strictly negative definite, so, $d\dot{d} < 0$, which is easily proven as long as $d \neq 0$.

3.2.1.2 *Generic control law for static/dynamic target reaching/tracking*

This second control law [Benzerrouk et al., 2014] dedicated to unicycles, is much more generic and will be used for several developments throughout this manuscript. The robot is controlled according to its center, i.e., $(l_1, l_2) = (0, 0)$ (cf. Figure 3.2(a)) and allows to the robot (x, y, θ) to track and reach, in a stable way, the target $T \equiv (x_T, y_T, \theta_T)$ (cf. Figure 3.2(a)). It is considered for this control law that the target has a linear velocity v_T and a kinematic model given by:

$$\begin{cases} \dot{x}_T = v_T.cos(\theta_T) \\ \dot{y}_T = v_T.sin(\theta_T). \end{cases} \tag{3.5}$$

The control law permitting us to track/reach, with asymptotic stability, the target is given by:

$$v = v_{max} - (v_{max} - v_T)e^{-(d^2/\sigma^2)} \tag{3.6a}$$

$$w = w_S + ke_\theta \tag{3.6b}$$

where:

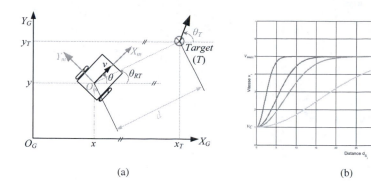

(a)   (b)

FIGURE 3.2 (a) Unicycle and target configuration in global and local reference frames. Control variables according to Lyapunov synthesis are also shown. (b) Evolution of the robot's linear velocity v according to the distance d and to the value of σ.

- v_{max} is the robot maximum linear velocity. Naturally, v_T has to be such that $v_T \leq v_{max}$.

- $d = \sqrt{(x - x_T)^2 + (y - y_T)^2}$ is the Euclidean distance between the robot center and the target to reach (cf. Figure 3.2(a)).

- e_θ is the orientation error such that $e_\theta = \theta_{T_{Setpoint}} - \theta$. With $\theta_{T_{Setpoint}}$ is given by [Benzerrouk et al., 2014]:

$$\theta_{T_{Setpoint}} = \arcsin(\frac{v_T}{v} \sin(\theta_T - \theta_{RT})) + \theta_{RT}. \tag{3.7}$$

- $w_S = \dot{\theta}_{T_{Setpoint}}$ and σ, k are positive constants permitting us to modulate the velocity of the robot's convergence toward the target T [Benzerrouk, 2010]. Figure 3.2(b) shows the influence of σ according to the robot's linear velocity.

This control law (cf. equations 3.6 (a) and (b)) allows to continually decrease, with asymptotic stability, the distance d and the angular error e_θ to zero in finite time [Benzerrouk et al., 2014]. The interesting aspect in this control law corresponds to the fact that it includes within it, specific angular set-points formulation (cf. equation 3.7) to insure the asymptotic stability of the robot (x, y, θ) toward its assigned target $(x_T, y_T, \theta_T, v_T)$. This control law gives thus a good balance between what should be defined as set-points and the control law formulation (cf. section 1.3.4, page 14).

3.2.2   Dedicated controller to tricycle mobile robots

Since the following proposed control law [Vilca et al., 2015a] is dedicated to a tricycle robot model [Luca et al., 1998], let us review below its well-known kinematics model (cf. equation 3.8). It is important to notice that the tricycle model given in

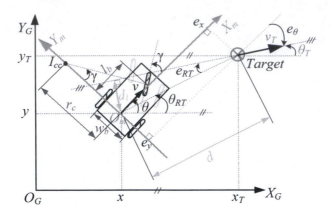

FIGURE 3.3 Tricycle and target configuration in global and local reference frames. Control variables according to Lyapunov synthesis are also shown.

equation 3.8 is used notably to model VIPALAB vehicles (cf. Annex A, page 185). Indeed, knowing that this experimental platform is devoted to urban transportation, this vehicle moves on asphalt with a low velocity (less than $3\ m/s$). Therefore, it appears quite natural to rely on a kinematic model, and to assume pure rolling and non-slipping at wheel–ground contact.

$$\begin{cases} \dot{x} &= v\cos(\theta) \\ \dot{y} &= v\sin(\theta) \\ \dot{\theta} &= v\tan(\gamma)/l_b \end{cases} \tag{3.8}$$

where (x, y, θ) is the posture (configuration state) of the robot at the point O_m (origin of the local reference frame $X_m Y_m$ linked to the robot (cf. Figure 3.3)), γ (the steering angle) corresponds to the orientation of the equivalent front wheel (cf. Figure 3.3), v is the linear velocity of the robot at O_m and l_b is the robot's wheelbase. According to Figure 3.3, w_b corresponds to the track width of the robot and I_{cc} the instantaneous center of curvature of the robot trajectory. The radius of curvature r_c is given by:

$$r_c = l_b/\tan(\gamma) \tag{3.9}$$

and $c_c = 1/r_c$ is the curvature of the robot trajectory. The minimal robot's curvature is defined by $r_{c_{min}} = l_b/\tan(\gamma_{max})$, where γ_{max} is the maximal robot's steering angle.

The proposed control law [Vilca et al., 2015a] aims to drive the robot toward specific targets (static or dynamic) in the environment. In order to have a self-contained section, let us give the main elements of synthesis, using a Lyapunov formulation, for this control law. At each sample time, the tracked target is defined by a posture (x_T, y_T, θ_T) and a velocity v_T. For genericity, the target is considered as a point with non-holonomic constraints (cf. Figure 3.3)), and its kinematic characteristics are given by:

$$\begin{cases} \dot{x}_T &= v_T \cos(\theta_T) \\ \dot{y}_T &= v_T \sin(\theta_T) \\ \dot{\theta}_T &= \omega_T. \end{cases} \quad (3.10)$$

where v_T and ω_T are respectively the linear and angular velocities of the target. Its radius of curvature is computed by $r_{c_T} = v_T/\omega_T$.

Before presenting the control law, let us introduce the control variables of the system (cf. Figure 3.3). The errors (e_x, e_y, e_θ) between the desired vehicle's pose (x_T, y_T, θ_T) and its current pose (x, y, θ) are given in the local reference frame $X_m Y_m$ by:

$$\begin{cases} e_x = & \cos(\theta)(x_T - x) & + \sin(\theta)(y_T - y) \\ e_y = & -\sin(\theta)(x_T - x) & + \cos(\theta)(y_T - y) \\ e_\theta = & \theta_T - \theta \end{cases} \quad (3.11)$$

A new error function e_{RT} is added to the canonical error system (3.11) (cf. Fig. 3.3). Let us first define the distance d and the angle θ_{RT} between the target and the vehicle's positions as:

$$d = \sqrt{(x_T - x)^2 + (y_T - y)^2} \quad (3.12)$$

$$\begin{cases} \theta_{RT} = & \arctan\left((y_T - y)/(x_T - x)\right) & \text{if } d \geq \xi \\ \theta_{RT} = & \theta_T & \text{if } d < \xi \end{cases} \quad (3.13)$$

where ξ is a small positive value ($\xi \approx 0$).

The error e_{RT} is related to the vehicle position (x, y) with respect to the target orientation (cf. Fig. 3.3). It is defined as:

$$e_{RT} = \theta_T - \theta_{RT}. \quad (3.14)$$

The adopted Lyapunov function V is given by equation (3.15) (cf. Figure 3.3):

$$\begin{aligned} V &= \frac{1}{2} K_d d^2 + \frac{1}{2} K_l d_l^2 + K_o[1 - \cos(e_\theta)] \\ &= \frac{1}{2} K_d d^2 + \frac{1}{2} K_l d^2 \sin^2(e_{RT}) + K_o[1 - \cos(e_\theta)]. \end{aligned} \quad (3.15)$$

The Lyapunov function (3.15) is therefore a function of three parameters which depend on the distance d between the target and the robot position; the distance d_l from the robot to the target line (line that passes through the target position with an orientation equal to the target orientation), this term is related to the Line of Sight and Flight of the target [Siouris, 2004]; and the orientation error e_θ between the robot and the target.

The desired linear velocity v and the front wheel orientation γ which permit us to asymptotically stabilize the error vector $(e_x, e_y, e_\theta, (v - v_T))$ toward zero (allowing us therefore to have $\dot{V} < 0$) are given by:

$$v = v_T \cos(e_\theta) + v_b \quad (3.16)$$

$$\gamma = \arctan(l_b c_c) \quad (3.17)$$

where v_b and c_c given by:

$$v_b = K_x \left[K_d e_x + K_l d \sin(e_{RT}) \sin(e_\theta) + K_o \sin(e_\theta) c_c \right] \tag{3.18}$$

with:

$$c_c = \frac{1}{r_{cT} \cos(e_\theta)} + \frac{d^2 K_l \sin(e_{RT}) \cos(e_{RT})}{r_{cT} K_o \sin(e_\theta) \cos(e_\theta)} + K_\theta \tan(e_\theta)$$
$$+ \frac{K_d e_y - K_l d \sin(e_{RT}) \cos(e_\theta)}{K_o \cos(e_\theta)} + \frac{K_{RT} \sin^2(e_{RT})}{\sin(e_\theta) \cos(e_\theta)} \tag{3.19}$$

and the initial values of e_{RT} and e_θ must satisfy the following initial conditions [Vilca et al., 2015a]:

$$e_{RT} \in]-\pi/2, \pi/2[\quad and \quad e_\theta \in]-\pi/2, \pi/2[. \tag{3.20}$$

$\mathbf{K} = (K_d, K_l, K_o, K_x, K_\theta, K_{RT})$ is a vector of positive constants defined by the designer. Accurate analysis of this stable and efficient control law is given in [Vilca et al., 2015a].

As it is shown in section 2.4 (page 37), different robots' sub-tasks could be described in a uniform way where the robot has to reach/follow/track specific target setpoints $(x_T, y_T, \theta_T, v_T)$. The presented control law will be used intensively through all this manuscript when the simulations or the experiments deal with vehicle control (it is notably used in chapters 4, 5 and 6).

It is important to mention that knowing the non-holonomy of the used robots (unicycle or tricycle) and their structural constraints (maximum linear and angular velocities), it is important to constrain the target dynamic to be always reachable by the robot's actual actuators. This important issue will be accurately addressed in chapter 6, where the proposed generic control laws are applied to control the formation of a group of robots (either unicycles (cf. section 6.3.3, page 149) or tricycles (cf. section 6.3.4, page 160)).

3.3 PROPOSED HYBRID$_{CD}$ CONTROL ARCHITECTURES

Once the stability of each elementary controller is proved (cf. section 3.2), let us put them in specific multi-controller architectures to perform complex tasks. The objective of the following proposed Hybrid$_{CD}$ (continuous / discrete) control architectures is to ensure in addition the stability and the smoothness of the overall control (cf. section 1.4.2, page 19). This can be done mainly if the coordination between the elementary controllers is mastered.

As given in section 1.4.1 (page 17) there exist two major principles of controller coordination: *action selection* and *fusion of actions*. Even if *fusion of actions* process gives very interesting robot behaviors, as it has been shown in [Adouane and Le-Fort-Piat, 2006] (using a kind of schema motor principle); in [Dafflon et al., 2015] (using Multi-Agent System) or in [Ider, 2009]

[Boufera et al., 2014] (using Fuzzy Logic principles), nevertheless the stability of the overall control architecture remains very complex, even impossible to demonstrate. However, control architectures based on the *action selection* process are relatively much easier to demonstrate even when switches between behaviors occur [Branicky, 1998] [Zefran and Burdick, 1998] [Liberzon, 2003]. However, the challenge is, in addition to the overall stability, to guarantee control smoothness. In fact, during mainly the phase of switching between controllers, the robot's setpoints or the control could be subject to jerking/discontinuities/oscillations (cf. section 1.4.2, page 19), the objective of the proposed Hybrid$_{CD}$ control architectures is therefore to avoid (or at least minimize) these drawbacks to obtain finally reliable and smooth robot navigation [Adouane, 2009a], [Benzerrouk et al., 2009], [Benzerrouk et al., 2010a] [Adouane, 2013].

The features of the presented Hybrid$_{CD}$ control architectures will be emphasized while performing on-line robot navigation in unknown and cluttered environments. The robot must thus discover its environment and act reactively to unexpected events (e.g., obstacle to avoid) while guaranteeing to reach its final target. In what follows, the focus will be on two proposed hybrid multi-controller architectures corresponding to the extension of the architectures given in section 2.5.1 (page 39). The objective is to add to these basic architectures, flexible and adaptive mechanisms of control (based on Adaptive Function (AF) (cf. section 3.3.1) or on Adaptive Gain (AG) (cf. section 3.3.2)) to guarantee at the same time, the overall control stability and the smoothness of the switch between controllers. Several simulations in cluttered environments permit us to confirm the reliability of the overall presented Hybrid$_{CD}$ architectures to obtain reliable and smooth robot navigation.

3.3.1   Hybrid$_{CD}$ CA based on adaptive function

The first mechanism of control [Adouane, 2009a], based on "Adaptive Function" (AF), permits us to manage the interaction between several controllers in a stable and smooth way. At the beginning, its global principles/concepts are presented (cf. subsection 3.3.1.1) before we apply it on an effective multi-controller architectures (cf. subsection 3.3.1.2). Furthermore, a specific "safety mode" is developed in subsection 3.3.1.2.3 to enhance the robot's safety. Several simulations in cluttered environments are given in subsection 3.3.1.3.

3.3.1.1   Global structure for AF

The blocks composing this first generic hybrid control architecture (cf. Figure 3.4) are detailed below before being applied to reactive robot navigation (cf. section 3.3.1.2).

Control law blocks   Every controller is characterized by a specific control law F_i, corresponding to a stable nominal law which is represented by the function:

$$F_i(S_i, t) = \eta_i(S_i, t) \tag{3.21}$$

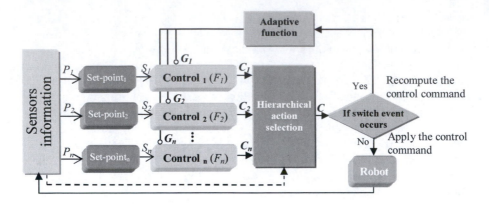

FIGURE 3.4 Generic Hybrid$_{CD}$ control architecture based on structure 1 (cf. Figure 2.7(a), page 39) and adaptive function.

where S_i is the set-point sent to the controller "i." Otherwise, in order to avoid the important controls jumps at the time, for instance, of the switch between controllers (e.g., from the controller "j" to the controller "i" at the instant t_0), an adaptation of the nominal law is proposed, F_i becomes thus:

$$F_i(S_i, t) = \eta_i(S_i, t) + G_i(S_i, t) \qquad (3.22)$$

with $G_i(S_i, t)$, the adaptive function (cf. equation 3.23), is a strict monotonous function that tends to zero after a certain amount of time "$T = H_i(P_i, S_i)$." The value of this time T depends on the criticality of the controller$_i$ to join as fast as possible the nominal control law $\eta_i(S_i, t)$. It is therefore an important part of the controller's "safety mode" (cf. subsection 3.3.1.2.3 for a specific example for obstacle avoidance controller).

$$G_i(S_i, t_0) = F_j(S_j, t_0 - \Delta t) - \eta_i(S_i, t_0) \qquad (3.23)$$

where Δt represents the sampling time between two control set-points and t_0 the time of abrupt change in a multi-controller's set-points.

The definition of $G_i(S_i, t)$ allows us to guarantee that the control law (cf. equation 3.22) tends toward the nominal control law after a certain time T, thus:

$$G_i(S_i, t \longrightarrow T) = \varepsilon \qquad (3.24)$$

where ε is a small constant value ≈ 0. The adaptive function $G_i(S_i, t)$ is updated every time a hard switch to the "i" controller occurs (cf. Figure 3.4).

The main challenge introduced by this kind of control is to guarantee the stability of the updated control law (cf. equation 3.22) even during the period where $|G_i(S_i, t)| \gg \varepsilon$.

Adaptive function block This block has as input the "conditional block" (cf. Figure 3.4) that verifies if a specific control switch event occurs. If it occurs, then it must update the "adaptive function" corresponding to the future active controller (cf. equation 3.23). The different configurations needing the activation of the adaptive function block are given when at least one of the following events occurs:

1. The "hierarchical action selection" block chooses to switch from one controller to another (cf. section 2.5.3 (page 42) for an example of such block for reactive navigation). In this case, the active controller at the current time "t" is different from the one activated at the "t-Δt" time.

2. An abrupt transition in the set-points S_i of the controller$_i$ is encountered. This case can happen even if the same controller is still active. For instance, it is enough that the "obstacle avoidance" controller chooses to avoid another obstacle (cf. Algorithm 2, page 43) or to switch from "attractive phase" to the "repulsive phase" (cf. Algorithm 3, page 45).

3.3.1.2 Application for reactive navigation

The generic control architectures based on adaptive function (AF) (cf. Figure 3.4) is applied in what follows for reactive navigation of a unicycle robot. This AF is applied on structure 1 (cf. Figure 3.5) [Adouane, 2009a] (each controller has its own control law), but could also be easily applied for structure 2 (common control law between controllers) [Adouane, 2013]. The details of the used controllers (set-points definition and the used control laws) are given below. The nominal control laws composing the architecture have been synthesized using the Lyapunov theorem, as given in section 3.2, and the modification of these control laws according to adaptive function mechanism are given in what follows.

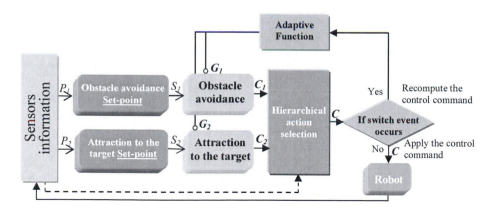

FIGURE 3.5 Dedicated Hybrid$_{CD}$ CA based on structure 1 and adaptive function (AF) for reactive navigation.

3.3.1.2.1 Attraction to the target controller with AF

As defined in section 2.5.4.2 (page 47), this controller has to attract the robot toward its final target $T_f = (x_f, y_f)$. It is obtained while using the nominal control law given in section 3.2.1.1; the homogeneous set-points inherent to this controller have to be defined simply with $T_{set-point} = (x_T, y_T, \theta_T, v_T) = (x_f, y_f, 0, 0)$.

The following show how to guarantee the right transition between controllers as described in section 3.3.1.1, and the modification of the nominal control law (cf. equation 3.3) while adding the term $G_A(t) = (G_{A\_v}, G_{A\_w})^T$ (cf. equation 3.22) will be highlighted. The control law becomes thus:

$$\begin{pmatrix} v \\ w \end{pmatrix} = -K \begin{pmatrix} \cos\theta & \sin\theta \\ -\sin\theta/_{l_1} & \cos\theta/_{l_1} \end{pmatrix} \begin{pmatrix} e_x \\ e_y \end{pmatrix} + \begin{pmatrix} G_{A\_v}(t) \\ G_{A\_w}(t) \end{pmatrix} \tag{3.25}$$

while considering the same Lyapunov function as given in equation 3.4. Therefore, $V_1 = \frac{1}{2}d^2$, with $d = \sqrt{e_x^2 + e_y^2}$ (distance robot-final target). The proposed new adaptive control law is asymptotically stable if $\dot{V}_1 < 0$. After some development it is deduced that to ensure this stability [Adouane, 2009a], K must be:

$$K > \frac{-(G_{A\_v}(t)e_x + G_{A\_w}(t)e_y)}{e_x^2 + e_y^2}. \tag{3.26}$$

As mentioned before, $G_{A\_v}(t)$ and $G_{A\_w}(t)$ functions must be chosen with respect to the constraints given in section 3.3.1.1. In fact, the absolute value of these functions must be monotonically decreasing according to the time "t," and they will be equal to zero after a certain time "T." Therefore, in order to always have bounded K, we must have $-(G_{A\_v}(t)e_x + G_{A\_w}(t)e_y) \leq e_x^2 + e_y^2$. Thus, to guarantee this assertion, it is sufficient to impose that $G_{A\_v}(t)$ decreases faster to zero than e_x and also that $G_{A\_w}(t)$ decreases faster to zero than e_y.

3.3.1.2.2 Reactive Obstacle avoidance controller with AF

As defined in section 2.5.4.1 (page 43), this controller has to reactively avoid any obstacle hindering the robot's movement toward its final target. The set-points necessary to this controller are defined in Algorithm 3, page 45). As a reminder, the used set-points correspond to $T_{set-point} = (x_T, y_T, \theta_T, v_T) = (x, y, \arctan(\frac{\dot{y}_s}{\dot{x}_s}), v_{const})$, where (x, y) correspond to the robot position; $\dot{x}_s$ and $\dot{y}_s$ are given by the differential equations describing the local planned PELC (cf. section 2.3.1, page 31); v_{const} corresponds to a constant linear velocity set-point. According to these set-point definitions, the used control is simplified therefore to an orientation control. Therefore, in terms of stable nominal control law, the one given in equation 3.6b will be used. The following shows how to guarantee the right transition between controllers as described in subsection 3.3.1.1, the modification of the nominal control law while adding the term $G_O(t)$ (cf. equation 3.27), it becomes thus:

$$w = w_S + ke_\theta + G_O(t) \tag{3.27}$$

where $G_O(t)$ the adaptive function. $\dot{e}_\theta = w - w_S$ is given then by:

$$\dot{e}_\theta = -ke_\theta - G_O(t) \tag{3.28}$$

while considering the same used Lyapunov function ($V_2 = \frac{1}{2}e_\theta^2$) to demonstrate the stability of the nominal control law (cf. section 3.2.1.2). The proposed new adaptive control law is asymptotically stable if $\dot{V}_2 < 0$. $\dot{V}_2$ is equal to $e_\theta\dot{e}_\theta = -ke_\theta^2 - G_O(t)e_\theta$. To guarantee the asymptotic stability of this new structure, k must verify:

$$k > -\frac{G_O(t)}{e_\theta} \tag{3.29}$$

where the $G_O(t)$ function is chosen with respect to the constraints given in section 3.3.1.1 and to the fact that it must decrease faster to zero than e_θ.

3.3.1.2.3 Obstacle avoidance safety mode

The adaptive function $G_O(t)$ applied to the obstacle avoidance controller permits us to obtain smooth robot control while maintaining the stability of the overall applied control law (cf. equation 3.27). However, during the time "T" (cf. section 3.3.1.1), the obstacle avoidance controller is far from its nominal law (given when $\mid G_O(t) \mid \gg \varepsilon$) and the robot can collide with obstacles. Thus, to ensure the smoothness of the control without neglecting the robot's safety, G_O will be parameterized according to the robot-obstacle distance "$d = D_{RO_i}$" (cf. Figure 2.2, page 27), G_O becomes:

$$G_O(t, d) = Ae^{Bt} \tag{3.30}$$

where:

- A value of the control difference between the control at the instants "$t - \Delta t$" and "t" (cf. equation 3.23),

- $B = log\left(\varepsilon/|A|\right)/T(d)$ with:

   - ε very small constant value ≈ 0 (cf. equation 3.24),

   - $\begin{cases} T(d) = T_{max} & \text{if } d > K_p \\ T(d) = c.d + e & \text{if } K_p \geq d \geq K_p - p.\underline{\text{Margin}} \\ T(d) = \varepsilon & \text{if } d < K_p - p.\underline{\text{Margin}} \end{cases}$

      ○ K_p corresponds to the safe distance to the Surrounded Ellipse (cf. section 2.3 and Figure 2.2 (page 27)) with $K_p = R_R + \underline{\text{Margin}}$,
      ○ p positive constant < 1 which allows us to adapt the maximum distance "$d = d_{Max}$" where the adaptive function must tend toward zero. As p becomes smaller, more priority is given to the safety behavior instead to the smoothness of the switch between controllers,
      ○ $c = T_{max}/p.\underline{\text{Margin}}$
      ○ $e = T_{max}(1 - K_p/p.\underline{\text{Margin}})$

Therefore, $T(d)$ goes from T_{max} to 0 while following a linear decrease. If the robot is at a distance bigger than K_p, then $T = T_{max}$ and decreases linearly to become 0 when $d < K_p - p.\underline{\text{Margin}}$. This function permits thus, when $d < K_p - p.\underline{\text{Margin}}$, to remove completely the effect of the adaptive control function and ensures the priority of the safety of the robot's navigation.

3.3.1.3 Simulation results

In this section, several simulations on different robot configurations and cluttered environments will permit us to confirm the reliability and the efficiency of the proposed control architectures based on adaptive functions. Figure 3.6(a) shows an example of obtained smooth robot trajectory when the proposed control architecture (cf. Figure 3.5) is used. It shows also clockwise and counter-clockwise obstacle avoidance using on-line set-point-based circular limit-cycles. Figure 3.7 shows respectively the evolution of v and w robot velocities when the adaptive functions are not used. These controls are much more abrupt than those obtained when the adaptive functions are used (cf. Figure 3.8).

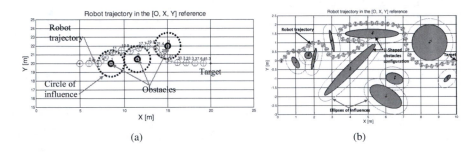

(a)   (b)

FIGURE 3.6 Smooth robot trajectories using the proposed control architecture based on adaptive function (a) and structure 1 [Adouane, 2009a] (b) and structure 2 [Adouane, 2013].

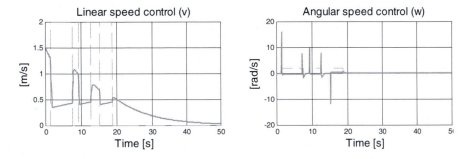

FIGURE 3.7 Control without adaptive mechanism.

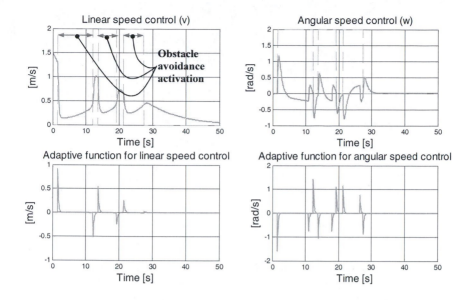

FIGURE 3.8   Control with adaptive mechanism.

Furthermore, to evaluate the actual robot's trajectory enhancement in terms of smoothness [Fleury et al., 1993], [Gulati, 2011], the criteria given by equations 3.31 and 3.32 are used:

$$I_v = \int_0^{T_T} |\dot{v}|dt \qquad (3.31)$$

and

$$I_w = \int_0^{T_T} |\dot{w}|dt \qquad (3.32)$$

where $\dot{v}$ and $\dot{w}$ correspond respectively to linear and angular robot acceleration, and T_T is the necessary time, for the robot, to reach the target. According to these two indicators, it was observed a significant gain in the smoothness of v and w equal to 6% and 50% respectively. Moreover, to better highlight the smoothness enhancement, a statistical survey was made in [Adouane, 2013] while doing a large number of simulations in different cluttered environments where for each one, a navigation with and without adaptive function is performed. We did specifically 1000 simulations with each of them, 10 obstacles with different random positions in the environment (cf. Figure 3.6(b) for an example of obtained trajectory with AF). A significant gain was observed in the smoothness of v and w controls that are equal respectively to 30% and 35%. Note that these last results were given using structure 2 control architecture (cf. Figure 3.14), where both controllers have the same control law [Adouane, 2013].

The simulation depicted in Figure 3.9 permits us to demonstrate the relevance of

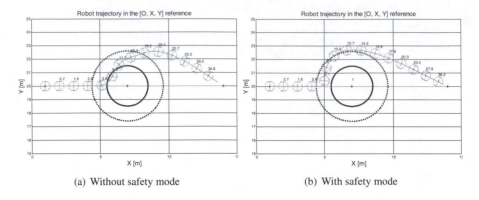

(a) Without safety mode (b) With safety mode

FIGURE 3.9 Robot trajectories without and with "safety mode."

the proposed safety mode (cf. section 3.3.1.2.3), especially when the robot navigates very close to obstacles. It shows the case where the obstacle avoidance controller applies or does not apply the safety mode. When it is not applied, the robot hit the obstacle (cf. Figure 3.9(a)).

Figure 3.10 gives the evolution of the adaptive function when the safety mode is applied (cf. Figure 3.10(b)) or not (cf. Figure 3.10(a)). It is observed in Figure 3.10(b) that the maximal time T_{max} to achieve the interpolation decreases every time that the robot moves dangerously closer to the obstacle. Figure 3.11 shows that the overall proposed structure of control is stable, and that the Lyapunov function of each controller $V_i|_{i=1..2}$ always decreases asymptotically to the equilibrium point even when the adaptive safety mode is applied.

3.3.2 Hybrid$_{CD}$ CA based on adaptive gain

The second presented mechanism of control, based on "adaptive gain" (AG) [Benzerrouk et al., 2010a] [Benzerrouk et al., 2010c], permits also to manage the interaction between several controllers in a stable and smooth way. At the beginning, its global principles/concepts are presented (cf. subsection 3.3.2.1) before to apply it on an effective multi-controller architectures (cf. subsection 3.3.2.2). Several simulations in cluttered environments are given in subsection 3.3.2.3.

3.3.2.1 Global structure for AG

As highlighted in section 1.4.2 (page 19), even if each controller is individually stable, it is important to constrain the switch between them to avoid instability of the overall system [Liberzon, 2003]. It is proposed in what follows to use the Multi-Lyapunov Function (MLF) [Branicky, 1998] as a theoretic background to demonstrate the stability of the proposed multi-controller architectures. Let us give first the definition of the Multiple Lyapunov function theorem.

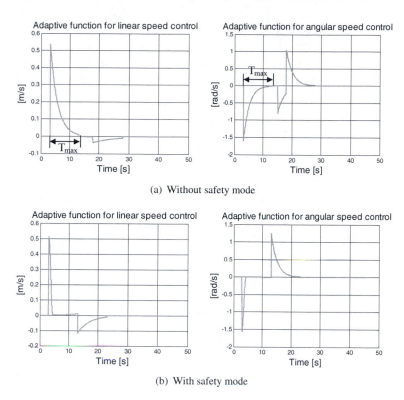

(a) Without safety mode

(b) With safety mode

FIGURE 3.10 Adaptive function evolution.

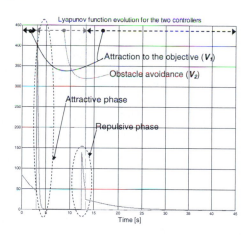

FIGURE 3.11 V_1 and V_2 Lyapunov functions evolution when the safety mode is used.

Theorem 3.1 *Multiple Lyapunov Functions (MLF)*
Given N dynamical subsystems $\sigma_1, \sigma_2, ... \sigma_N$ (where each of them has an equilibrium point at the origin), and N candidate Lyapunov functions, $V_{\sigma_1}, V_{\sigma_2}, ..., V_{\sigma_N}$. For each subsystem σ_i, let $t_1, t_2, ..., t_m, ..., t_k$ be the switching moments to this subsystem (only one subsystem is active at a time).
If always (V_{σ_i} decreases when σ_i is active) and ($V_{\sigma_i}(t_m) < V_{\sigma_i}(t_{m-1})$)
Then *the overall hybrid system is Lyapunov asymptotic stable.*

This theorem is illustrated in Figure 3.12 for the elementary sub-system σ_i. When σ_i is active (phases I and III), its Lyapunov function decreases. When the control switches to another subsystem (phases II and IV), V_{σ_i} may increase. However, to ensure the global stability according to MLF theorem, this subsystem (as for all other subsystems) must be reactivated only if its Lyapunov function takes a smaller value than the last time the system switches in. In the example depicted in Figure 3.12, this corresponds to having $V_{\sigma_i}(t_k) < V_{\sigma_i}(t_{k-1})$.

According to this theorem, it appears that V_{σ_i} could increase when it is not active but the condition which must be verified for all the sub-systems allows us to ensure the overall stability of the hybrid system. This definition corresponds to a kind of weak stability [Brogliato et al., 1997] [Liberzon, 2003].

In the basic used multi-controller architectures (cf. section 2.5.1, page 39), during the critical phase of switching between controllers, the different used Lyapunov functions can increase. This happens inevitably due to the discontinuity of the different controllers set-points or to the heterogeneity of the used Lyapunov functions (control laws). To demonstrate the overall stability of the proposed architectures, the MLF theorem has been used in [Benzerrouk et al., 2008] and [Benzerrouk et al., 2009] to control a mobile robot following a trajectory in the presence of obstacles. The basic used multi-controller architecture is composed of two elementary controllers ("trajectory tracking" and "obstacle avoidance"), each of them has its own control law. To satisfy the MLF theorem, a third controller ("go to the goal") has been added to the initial architecture. Nevertheless, the obtained control architecture is not suitable for highly cluttered environments since it imposes a lot of constraints. In fact, due to the

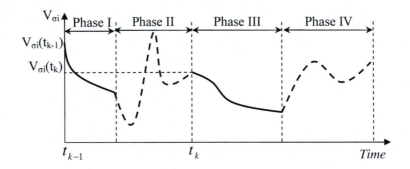

FIGURE 3.12 Variation of the Lyapunov function for the σ_i subsystem. Solid lines indicate that σ_i is active and dashed for inactive.

unknown nature of the environment, it is not always possible to let active a controller as long as it is necessary to respect the MLF theorem, mainly, if the robot safety is not ensured.

The following shows the overall principle to ensure the MLF stability while taking into account the robot's integrity. To achieve that, it has been proposed in [Benzerrouk et al., 2010a] [Benzerrouk et al., 2010c] [Benzerrouk, 2010] to act on the convergence rate of the used Lyapunov functions. Indeed, the switching from one controller σ_j to another σ_i must happen only if (cf. Figure 3.13):

$$V_{\sigma_i}(t_s + \tau) < V_{\sigma_i}(t_{bes}) \qquad (3.33)$$

where $V_{\sigma_i}(t_{bes})$ is the value of V_{σ_i} at the time preceding the last switching from σ_i to another controller (cf. Figure 3.13); t_s corresponds to the time of switching; τ corresponds to the time taken by the Lyapunov function to satisfy equation 3.33.

According to the MLF theorem, when a stable controller is active, it must remain at least for a time τ to ensure the asymptotic stability of the overall hybrid system. It is proposed in what follows to adapt the used control laws to ensure the satisfaction of equation 3.33. The value of τ must be appropriately chosen in order to take into account the robot's structural constraints and the navigation safety.

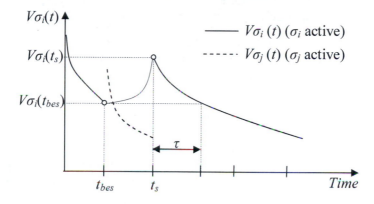

FIGURE 3.13 Values of $V_{\sigma_i}(t_{bes})$ and $V_{\sigma_i}(t_s)$ corresponding to the system σ_i.

3.3.2.2 Application for dynamic target tracking in a cluttered environment

In what follows it is proposed to use the Multi-Lyapunov Functions (cf. Theorem 3.1) and the constraints on τ (given according to equation 3.33) in order to ensure the overall stability of the proposed multi-controller architecture (cf. Figure 3.14). This architecture uses a single control law for both controllers ("attraction to the target" and "obstacle avoidance") to achieve their respective sub-tasks. The used stable control law is given by equations (3.6a) and (3.6b).

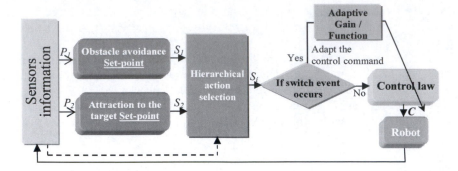

FIGURE 3.14 Dedicated Hybrid$_{CD}$ CA based on structure 2 (cf. section 2.5.1, page 39) and adaptive gain for reactive navigation.

The set-point definition $T = (x_T, y_T, \theta_T, v_T)$ for each of these controllers is summarized as follows:

- For "Attraction to the target" $T_{AT} = (x_T, y_T, \theta_{T_{AT}} = \arctan(\frac{y_T - y}{x_T - x}), 0)$, where (x, y, θ) correspond to the robot's pose and (x_T, y_T) the current target position,

- For 'Obstacle avoidance" $T_{OA} = (x, y, \theta_{T_{OA}} = \arctan(\frac{\dot{y}_s}{\dot{x}_s}), v_{const})$, where $(\dot{x}_s, \dot{y}_s)$ correspond to the generated PELC for reactive obstacle avoidance (cf. Algorithm 3, page 45).

The following analysis will be made on the robot's angular velocity (cf. equation 3.6b). The angular error e_θ is given by $e_\theta = \theta_T - \theta$ where $\theta_T = \theta_{T_{AT}}$ or $= \theta_{T_{OA}}$ according to the active controller. Therefore, at each controller switch a set-point discontinuity occurs, but this discontinuity can occur for different other events such as an abrupt change in the dynamic of the target to each.

The angular control has been proved stable (cf. section 3.2.1.2) while using the Lyapunov function:

$$V = \frac{1}{2} e_\theta^2. \tag{3.34}$$

It is simple to prove the stability of the used control law, since the error e_θ have the following dynamic:

$$e_\theta(t) = e_\theta(t_s) e^{-k(t - t_s)} \tag{3.35}$$

with $e_\theta(t_s)$ the value of angular error at the commutation time t_s (corresponding to the initial time to the new active controller).

The Lyapunov function evolution will be then:

$$V(t) = (e_\theta^2(t_s)/2) e^{-2k(t - t_s)} \tag{3.36a}$$
$$V(t) = V(t_s) e^{-2k(t - t_s)}. \tag{3.36b}$$

From equation 3.36b, it is possible to compute $V(t_s + \tau)$:

$$V(t_s + \tau) = V(t_s)e^{-2k\tau}. \tag{3.37}$$

To find the minimal time τ, during which the controller must stay active before the next switch, and permitting us to guarantee the asymptotic stability of the hybrid system, τ is computed as [Benzerrouk et al., 2010c]:

$$\tau > \frac{ln(V(t_{bes})/V(t_s))}{-2k}. \tag{3.38}$$

It is observed clearly that the time τ depends on the gain k of the control law. Indeed, this one is related to the velocity of angular error convergence, and therefore to the Lyapunov function (cf. equation 3.36).

However, the gain k must always comply with the structural limitations of the robot (mainly, according to its maximal angular velocity w_{max}). After some developments given in [Benzerrouk, 2010, chapter 3], authorized k_{max} is obtained:

$$k_{max} = \frac{\lambda\pi}{|e_\theta(t_s)|} \tag{3.39}$$

where λ corresponds to a positive real value determined while taking into account the robot's maximal angular acceleration $\dot{w}_{max}$.

Thus, according to this maximal value k_{max}, it is possible to minimize the time τ in order to reduce any latency of the effective activation of the next controllers. The minimum possible value is:

$$\tau_{min} > \frac{ln(V(t_{bes})/V(t_s))}{-2k_{max}} \tag{3.40}$$

provided that $V(t_{bes}) \neq 0$.

Once the time τ_{min} elapses, if there is no switching, k gain should recover its initial suitable value k_{ini} in a smooth way. Knowing that e_θ always decreases exponentially, it is proposed that k follows this dynamic:

$$k = k_{ini} - (k_{ini} - k_{max})\tanh^2(e_\theta). \tag{3.41}$$

Obviously, when the activation of a critical controller is mandatory (an obstacle avoidance, for instance), even if the time τ_{min} is not elapsed, this controller must be activated instantaneously while having $k = k_{max}$ (therefore, with the maximal convergence rate of the Lyapunov function).

3.3.2.3 Simulation results

To highlight the efficiency of the proposed mechanism of control, based on adaptive gain (AG) to ensure the overall control stability, some simulations are shown below. The multi-controller architecture depicted in Figure 3.14 is used to perform a dynamic target tracking in a cluttered environment.

Figures 3.15(a) and 3.15(b) emphasize the aspect that even if the robot succeeds

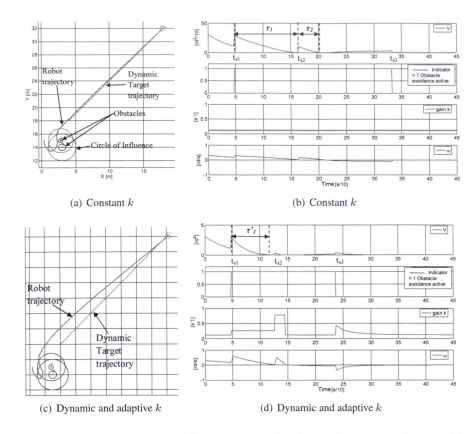

(a) Constant k

(b) Constant k

(c) Dynamic and adaptive k

(d) Dynamic and adaptive k

FIGURE 3.15 Importance of the proposed adaptive gain process. (a) and (c) Robot's trajectories while using, respectively, constant gain k or a dynamic adaptive gain. (b) and (d) Lyapunov function V and the angular velocity ω evolution without and with dynamic gain, respectively.

in reaching the dynamic target, the constant imposed value of $k = k_{ini}$ does not permit us to increase the reactivity of the robot to avoid more safely the obstacles' configuration. According to MLF theorem, the system is nevertheless asymptotically stable.

While taking the same initial navigation conditions, Figures 3.15(c) and 3.15(d) show the case where the gain k is variable defined initially according to equation 3.40. It is observed in this simulation that the first convergence time τ_1' of V (cf. Figure 3.15(d)) is smaller than τ_1 when the gain is constant (cf. Figure 3.15(b)). The obstacle is therefore avoided more safely while respecting the MLF theorem. It is observed also in this last simulation that the gain k get back to its initial value k_{ini} whenever the effect of switching is dissipated. Finally, let us note also that the angular robot velocity ω varies more significantly than in the case of a constant gain but while remaining unsaturated (maximal angular acceleration was taken $= 1rd/s^2$) [Benzerrouk, 2010, chapter 3].

3.4 CONCLUSION

This chapter focused on stability, according mainly to Lyapunov synthesis, for elementary proposed controllers as well as for the overall Hybrid$_{CD}$ (continuous/discrete) multi-controller architectures.

First, the main proposed control laws to reach/track a static/dynamic target were presented. These control laws permit us to control in a stable and flexible way unicycle robots [Benzerrouk et al., 2014] as well as tricycle robots [Vilca et al., 2015a]. These developments are complementary with those of chapter 2, where a homogeneous set-points definition were detailed. In fact, having appropriate set-points definition and generic stable control laws to stabilize the error to zero allow us to lead at the end to highly reliable controllers to achieve several sub-tasks. These controllers will be used in the following chapters to exhibit intelligent mono and multi-robot navigation.

The other part of this chapter has been dedicated to proving the overall stability and smoothness of the developed multi-controller architectures (even at the critical switching moments). After giving a short overview about the different proposed techniques to manage the coordination of several elementary controllers, the potentialities of Hybrid$_{CD}$ systems have been taken as a formal framework to demonstrate the overall stability of such architectures based on "Action selection." In this kind of controller coordination, two mechanisms of control, based on adaptive function (cf. section 3.3.1) or adaptive gain (cf. section 3.3.2) have been presented and implemented on effective architectures. These architectures aimed to achieve on-line, smooth and safe robot navigation in cluttered environments. Several simulations were shown to highlight the effectiveness of the different proposals.

Fully autonomous robot navigation must show accurate demonstration of the reliability of its different components and their interactions (cf. section 1.2, page 7). The control mechanisms given in this chapter contribute, even with small increments, to this important challenging issue.

CHAPTER 4

Hybrid$_{RC}$ (reactive/cognitive) and homogeneous control architecture based on PELC

CONTENTS

T HIS CHAPTER focuses on the aptitude of a control architecture to exhibit reactive as well as cognitive skills according to the robot's situation. The main objective of this kind of Hybrid$_{RC}$ (Reactive/Cognitive) structure is to enable us to deal on-line and safely with unpredictable or uncertain situations, and to optimize the overall robot navigation if the environment is well-known/mastered (e.g., accurate localization, large perception field with low uncertainty, etc.). Furthermore, since the previous chapters have emphasized reactive navigation, this chapter will pay more attention to the proposed planning methods, mainly the one based on PELC for a car-like robot. Afterward, a homogeneous (in term of set-points and control law) and $Hybrid_{RC}$ multi-controller architecture will be presented in detail.

4.1 HYBRID$_{RC}$ CONTROL ARCHITECTURES

After highlighting in section 1.3.2 (page 11) what is meant by reactive and cognitive robot behaviors, let us present in what follows an overview of the architectures showing the two behaviors. More and more control architectures exhibiting $Hybrid_{RC}$ (Reactive/Cognitive) structure appear in the literature. This enables us to keep only the advantages of the two structures while minimizing their drawbacks [Firby, 1987], [Arkin, 1989a], [Gat, 1992] [Ranganathan and Koenig, 2003] [Alami et al., 1998] [Ridao et al., 1999] [Grassi Junior et al., 2006] [Rouff and Hinchey, 2011]. Several $hybrid_{RC}$ control architectures have been explored in the literature. An interesting survey of 22 control architectures has been given in [Ridao et al., 1999], highlighting among other things, how developments for Unmanned Grounded Vehicle (UGV) architectures have been extended for autonomous underwater vehicles (the addition of the 3D navigation and the specificity of the sea environment). Usually in the literature, a consensus is adopted for the structure of hybrid control architectures which is generally structured in three layers: the highest level is responsible for mission planning and re-planning; the intermediate layer activates the low-level behaviors and permits passing of parameters to them; while the lowest layer level (called commonly the reactive layer) contains the physical sensors and actuator interfaces. The cognitive part (highest level) contains generally a symbolic world model (based on artificial intelligence concepts), which develops plans and makes decisions on the way to perform the robot's objectives. The more reactive part (the two other lower levels) are responsible for reacting to local events without complex reasoning. Nevertheless, generally the structural conception of these hybrid architectures remain too complex to manage the different level of hierarchy imposed by this kind of architecture [Gat, 1998] [Arkin, 1989a]. They are also low homogenized to deal with the effective set-points to send to the robot's actuators (lowest level). Efforts have been concentrated on the conceptual aspects (using for instance the multi-agent paradigm to manage the multi-layered proposed architectures [Konolige et al., 1997] [Busquets et al., 2003] [Hsu and Liu, 2007]) and less on the overall control simplicity, genericity and its effective implementation [El Jalaoui et al., 2005] [Mouad et al., 2012]. Indeed, even if the control architecture must show a good level of knowledge abstraction and decision, it is important also to translate these aspects in terms of low-level vehicle control to exhibit clearly its effects on the vehicle movements, which permit at its turn to attest to the safety and the overall stability of the control architecture [Adouane, 2013].

As shown in the previous chapters, the robot had to navigate mainly while using reactive characteristics (cf. section 1.3.2, page 11). Even if the robot's features have been satisfactory to perform reliable navigation, to permit exhibiting much more cognitive characteristics (cf. section 1.3.2, page 11) it is important to have appropriate techniques to define its future tasks/movements while integrating more environmental knowledge. The aim is obviously to optimize its long-term navigation aspects. The planning phase is therefore, in the targeted robot tasks, to attest to the optimality of the navigation. This is the main reason which led us to develop several techniques of planning. Furthermore, this chapter's main objective is presenting a proposed Homogeneous and Hybrid$_{RC}$ (reactive/cognitive) Control Architecture (HHCA) for vehicles navigating in different kinds of environments. This architecture enables us to manage simply the activation of reactive or cognitive navigation according to the environment context (uncertain or not, dynamic or not, etc.). This architecture is based, among other things, on the use of a homogeneous set-points definition and on an appropriate control law shared by all the controllers (composing the architecture).

This chapter is organized as follows. In section 4.2, a short overview of the different proposed planning techniques is given. A focus will be made in section 4.3 on one proposed planning method for local and global path optimization based on PELC (cf. section 2.3.1, page 31). Section 4.4 gives the details and the specificities of the proposed Hybrid$_{RC}$ control architecture while presenting its different constituting modules. An intensive validation by simulation of its different features will be given in section 4.4.4. This chapter ends with some conclusions and prospects.

4.2 OVERVIEW OF DEVELOPED PLANNING METHODS

As given in section 1.5 (page 21), several techniques of navigation exist. The majority of them are based on short or long-term path planning. In the first case, these techniques could be easily used for reactive navigations (cf. sections 1.3.2 (page 11) and 2.4.2 (page 38)). During our different works, several techniques of planning have been investigated (for both short-term and long-term planning), among the most important let us cite:

- **Artificial Potential Field** (APF): This well-known technique is among the most disseminated works in the literature due mainly to its simplicity and intuitive use. APF has been notably used in [Mouad et al., 2012] for planning and re-planning robot trajectories in cluttered and dynamic environments. Furthermore, these APF planning techniques have been embedded in an overall proposed control architecture, called MAS2CAR (Multi-Agent System to Control and Coordinate teAmworking Robots, cf. Annex A), as a decisional tool to manage/organize the activity of a group of robots[1] [Mouad et al., 2010] [Mouad et al., 2011a] [Mouad et al., 2011b].

- **Clothoid curves**: The Clothoid curves[2] generation have been investigated to generate smooth paths for autonomous navigation of vehicles

---

[1]MAS2CAR architecture permitted notably to achieve complex decision-making process, using organizational process-based on MAS to define which trajectory to take by each robot in the context of multi-robot / multi-target task.

[2]These curves permit smooth curvature variation according to the path length (curvilinear abscissa)

[Gim et al., 2014a] [Gim et al., 2014b]. Knowing that the analytic formulation of these curves is complex and their resolution for any initial and final vehicle's pose is not trivial, and has been investigated in these works to propose systematic iterative methodology for the generation of such paths. The obtained techniques are used as a local planner (for dynamic obstacle avoidance for instance [Gim et al., 2014b]) or as a global planner while connecting several elementary computed Clothoids to obtain the overall vehicle path [Gim et al., 2014a]. Several works are underway concerning the use of these developments on actual vehicles: e.g., for parking tasks (static use) or for smooth obstacle avoidance (dynamic use).

- **Multi-criteria optimization**: Two aspects have been addressed:

  - For waypoints generation: A navigation technique based on reaching sequential waypoints has been developed during our works (cf. chapter 5). To perform this kind of navigation, optimal techniques have been proposed to optimize the generation of a set of waypoints (number, poses, etc.). Specifically an Optimal Multi-criteria Waypoint Selection based on Expanding Tree (OMWS-ET) or based on Grid-Map (OMWS-GM) have been proposed (cf. section 5.4 (page 118) for more details).

  - For path planning-based PELC: The proposed techniques will be detailed in section 4.3.1 for optimal elementary PELC* and in section 4.3.2 for global path planning-based PELC (gPELC*).

4.3   OPTIMAL PATH GENERATION BASED ON PELC

It is proposed in what follows to obtain a generic way to enhance the use of the already presented PELC (cf. section 2.3, page 31), to perform optimally local obstacle avoidance as well as global robot navigation in cluttered environments. These optimized components (PELC* and gPELC*) will be afterward integrated in a Hybrid$_{RC}$ multi-controller architecture and will constitute an homogeneous way to obtain the vehicle's set-points.

4.3.1   Local path generation based on PELC*

The optimization of the already defined PELC path comes from notably the need to have a path which takes into account the robot's structural constraints (nonholonomy, maximal angular velocity, etc.). As an example, it is shown in Figure 4.1, the tracks of "PELC planned path" (the green discontinuous lines), which are not really followed by the robot, in fact, at each sample time, the robot computes the new control set-points given by an equation (cf. section 2.3, page 32). The showed PELC planned track corresponds to the limit-cycle path obtained the first time that the robot sees

---

[Walton and Meek, 2005]. Its use for vehicle path planning is interesting notably because the vehicle's steering angle can be simply defined w.r.t. these curves.

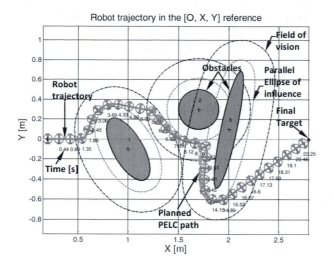

FIGURE 4.1 Reactive navigation in a cluttered environment using PELC. The green discontinuous lines correspond to initial computed PELC, when the current most obstructing obstacle is seen the first time.

the obstacle to avoid, this PELC does not take into account the robot's constraints. In addition, it is aimed through the optimization of PELC to choose effectively the robot initial pose (notably its initial orientation) and the obstacles' parameters (position, orientation, dimension, etc.) to find PELC* permitting to avoid optimally any obstacle.

Thus, while knowing the features of the Surrounded Ellipse (cf. Figure 2.2, page 27) (i.e., h, k, Ω, A and B) and the desired offset (safe distance K_p (cf. equation 2.3, page 32)) to the obstacle, it is shown in what follows how to obtain the optimal value of μ^* which enables us to minimize a multi-criteria function. This latter, called J (cf. equation 4.1), gathers different important sub-criteria linked to the features of the corresponding PELC path. Indeed, according to the value of μ in the PELC equation (cf. equation 2.3, page 32) the shape of the obtained limit-cycle can converge quickly or not to the assigned PEI (cf. Figure 2.5(a), page 34), but what is the optimal value of μ^* permitting to minimize the multi-criteria J? It is important to note that J evaluates the obtained PELC path while taking into account the initial robot configuration and its final reached configuration (which corresponds in what follows to the PELC configuration when it reaches the axis Y_{OT} (for an obstacle) or Y_T (for a target) (cf. Figure 2.5(a), page 34)).

$$J = w_1 J_{\text{DistanceToPTi}} + (1 - w_1) J_{\text{PELC}_{\text{Length}}} + w_2 J_{\text{PELC}_{\text{Curvature}}}$$
$$+ w_3 Bool_{\text{MaximumCurvature}} + w_4 Bool_{\text{Collision}}$$

$$(4.1)$$

where:

- $w_i \mid i = 1..4 \in \mathbb{R}^+$ are constants which permit the right balance between the different sub-criteria characterizing the computed PELC. The criteria J is defined therefore according to the vector $W_{PELC} = \{w_1, w_2, w_3, w_4\}$. It is to be noted that the weight $w_1 \in [0\ 1]$ and permits, as given in equation 4.1, balance between the values of $J_{\text{DistanceToPTi}}$ and $J_{\text{PELC}_{\text{Length}}}$ sub-criteria.

- $J_{\text{DistanceToPTi}}$ corresponds to the distance between the final reached position of the computed PELC and one among the points $PT_i \mid i = 1..2$ (chosen according to whether the PELC is obtained for clockwise or counter-clockwise avoidance (cf. equation 2.3, page 32)). For instance, Figure 2.5(a) (page 34) shows the value of $J_{\text{DistanceToPTi=2}}$ for a counter-clockwise PELC with $\mu = 0.1$.

- $J_{\text{PELC}_{\text{Length}}}$ corresponds to the curvilinear length of the obtained PELC. It is computed using the following equation.

$$J_{\text{PELC}_{\text{Length}}} = \int_{s_0}^{s_f} ds \qquad (4.2)$$

where s_0 and s_f correspond respectively to initial and final curvilinear abscissa of the obtained PELC.

- $J_{\text{PELC}_{\text{Curvature}}}$ characterizes the PELC curvature along its length. It is computed using the following equation.

$$J_{\text{PELC}_{\text{Curvature}}} = \int_{s_0}^{s_f} \mathcal{C}(s)^2\, ds \qquad (4.3)$$

where $\mathcal{C}(s)$ is the curvature at the abscissa s, $\mathcal{C}(s) = 1/\rho(s)$ with $\rho(s)$ the radius of curvature at s PELC abscissa.

- $Bool_{\text{MaximumCurvature}}$ corresponds to a Boolean value, which is equal to 1 if the maximum possible robot curvature is reached (or exceeded). It is to be noted that the weight w_3, linked to this sub-criteria, is a big positive value so that the overall criteria J will be highly penalized if the obtained PELC has at least one configuration where the robot must attain its maximum curvature (maximum steering angle) [Gim et al., 2014a].

- $Bool_{\text{Collision}}$ corresponds to a Boolean value, which is equal to 1 if the robot collides with at least one Surrounded Ellipse in the environment (cf. Figure 2.2, page 27). It is to be noted that the weight w_4, linked to this sub-criteria, is a very big positive value $\longrightarrow \infty$ so that the overall criteria $J \longrightarrow \infty$ if any obstacle is collided.

To obtain μ^* optimizing the PELC (cf. equation 2.3, page 32) according to the criteria J, this parametric optimization $\partial J/\partial \mu = 0$ should be computed. The mathematical formulation of this problem is highly non-linear, and the analytic solution is therefore complex to get. A numerical optimization is used to obtain μ^* (while using dichotomy for instance).

An important assumption in the proposed optimization is the fact that the robot, to perform its navigation, deals sequentially with only one obstacle/target at a time, until reaching its corresponding axis Y_{OT} (or Y_T), and switching to the other obstacle/target and so on until reaching the Y_T axis of the final target. In each elementary optimization, the obtained PELC* will permit either an obstacle avoidance behavior or an attraction toward a target. In these two optimizations, the parameters of the obtained PELC* are respectively:

- PELC*$((h,k), \Omega, (A, B, K_p), \mu^*)$, where h, k, Ω, A and B are the features of the detected obstacle and K_p the desired safe distance to this obstacle.

- PELC*$((x_f, y_f), 0, (\xi, \xi, \xi), \mu^*)$, where (x_f, y_f) is the position of the final target to reach and ξ is a very small value $\longrightarrow 0$.

These two elementary behaviors will be used to deal with different environments (cf. section 4.4.4) while performing either reactive or cognitive navigation, but first, let us introduce the proposed methodology to obtain long-term vehicle planning while using a sequence of appropriate PELC.

4.3.2   Global path generation based on gPELC*

This section will focus on the way to obtain optimal global path planning (from the robot initial configuration to a final desired configuration). In what follows, a multitude of PELCs are appropriately sequenced in one proposed algorithm to perform optimal global planning-based PELC (gPELC*). The proposed planning-based gPELC*, as the RRT* technique [Lavalle, 1998] [Karaman and Frazzoli, 2011], uses a tree of admissible paths (i.e., without collisions) while using elementary PELC, but contrary to RRT*, each individual path (branch) is, in addition to being safe, also the most suitable (in certain cases, the optimal one (cf. section 4.3.1)) to avoid each obstacle. The obtained overall path-based gPELC* is therefore closer to the effective optimal path leading the robot toward its final destination. The computation time is also much smaller w.r.t. RRT*, mainly due to the reasoning proposed for the tree expansion (cf. Algorithm 4), whereas the tree in RRT* is only expanded (randomly) by a constant length, depending on the adopted constant robot's velocities and $\delta T = t_{exp}$ (cf. section 5.4.1, page 118).

The proposed gPELC* has as objective to lead the vehicle from its initial pose $P_0 = (x_0, y_0, \theta_0, \gamma_0)$ (cf. the used vehicle model given in section 3.2.2 (page 57)) to a final assigned pose $P_f = (x_f, y_f, \Xi, \Xi)$, where (x_f, y_f) corresponds to the position of the final target and the Ξ symbol means here, any real value. Indeed, in this chapter, the values of the final vehicle's heading θ_f and front wheels angle γ_f are not imposed (cf. equation 3.8, page 58). The optimal methodology aims to connect several PELCs to reach P_f while allowing us to guarantee the safety and the smoothness of the obtained global path-based PELC (called gPELC). The targeted smooth path will enable us to generate smooth set-points for the control law, allowing thus to avoid the actuators jerking which ensure hence the passengers' comfort and preserves the actuators' lifetime [Gulati, 2011]. Obviously, the aim of the proposed optimal methodology is to ensure also the continuity of the vehicle heading θ and the wheels orientation γ (cf. Figure 3.3, page 58), even at the connection point between the PELCs.

To formalize the optimal path planning using a sequential concatenation of PELCs (gPELC*), let us use Graph Theory [Harary, 1969] [Bondy and Murty, 2008], through a shortest path-problem to optimize the several possible gPELCs. In graph theory, the shortest-path problem corresponds to finding a path between any two vertices (or nodes) in a graph such that the sum of the weights of its constituent edges is minimized. It is shown in section 4.3.1 that according to the value of μ the shape of the obtained PELC changes (cf. Figure 2.5(a)) and consequentially the value of J (cf. equation 4.1). The main idea is therefore to obtain the sequence of elementary PELC (with appropriate values of μ and direction (clockwise or not, according to the value of r in equation 2.3 (page 32)) which permit us to minimize the sum of costs leading the vehicle from P_0 to P_f.

It is supposed in what follows the presence of N obstacles in the environment, each one has an identification number id and is surrounded by an appropriate Parallel Ellipse of Influence (PEI$_{id}$, cf. Figure 2.2 (page 27)). PEI$_{id}$ is characterized by $[(h_{id}, k_{id}), \Omega_{id}, (A_{id}, B_{id}, K_{p\,id})]$ (cf. equation 2.1 (page 28) and section 2.3.1 (page 31)). id corresponds to the identifier of the obstacle $id = \{1, ...N\}$ or to $id = f$ if the PEI is linked to the final target. For easy understanding of the proposed overall optimal planning given in Algorithm 4, some definitions / conventions will be given below to formalize the optimization problem, using graph theory, more specifically while using a Tree structure:

- A Tree (T) is a directed rooted Graph in which any two vertices are connected by exactly one path (without closed loops (cycles)). The tree T is characterized by T = (V, E), where V and E are respectively the set of all vertices and all the edges of the obtained T.

- Each vertex $v_j \in$ V is characterized by a state $v_j \equiv [(x_j, y_j, \theta_j, \gamma_j),$ Parent(v_j)] = $[P_j, v_i]$ (where P_j corresponds to the vehicle's set-point when it will reach the vertex v_j). The tree root v_0 does not have a parent, $v_0 \equiv [P_0, 0]$. The vertex v_0 alone corresponds to level 0 (**Level$_0$**) of T and is characterized by a set of vertices represented by $S_0 = \{v_0\}$. S_i will correspond therefore to **Level$_i$** of the tree and will contain all the children vertices generated from vertices of S_{i-1} set. Each vertex $v_j \neq v_0$:

  - holds, as given above, the value of its parent v_i in T. We can write thus v_i = Parent(v_j) and v_j = Child(v_i), v_i and v_j are adjacent vertices.

  - contains the final state of one PELC$_i^j$ (when first reaching the Y axis of the reference frame ∞v_j). The symbol "∞" expresses the fact that the considered vertex v_j is defined w.r.t. the reference frame linked to one obstacle or to the final target (cf. section 2.3.2). In that case, it is written: $v_j \infty \mathfrak{R}_{id}$.

- Each edge $e_i^j \in$ E is characterized by a state $e_i^j \equiv [$ PELC$_i^j, J_i^j]$, where PELC$_i^j \equiv$ PELC$_i^j$(PEI$_{id}, r, \mu$) corresponds to a Parallel Elliptic Limit-Cycle linking the vertex v_i to $v_j \otimes \Re_{id}$. This PELC$_i^j$ has as the initial state, the posture defined in v_i and as final state, the posture defined in v_j. PELC$_i^j$ is characterized also by r and μ which correspond respectively to the PELC direction and specific shape (cf. section 2.3.1 (page 31) and Figure 2.5(a) (page 34)) which enable us to reach v_j from v_i (cf. equation 2.3, page 32). The edge e_i^j has a weight $J_i^j \equiv J_i^j(\text{PELC}_i^j)$ corresponding to the PELC$_i^j$ cost given by equation 4.1.

- The number of children (or also growing branches/edges) from each vertex v_i is fixed in the proposed Algorithm 4 through the pre-fixed constant $m \in \mathbb{N}^+$. This algorithm proposes to generate m PELC in each direction (clockwise and counter-clockwise, i.e., r will be equal to ± 1 respectively in equation 2.3 (page 32)). In each direction, several PELCs are generated for each μ value given in a predefined set $S_\mu = \{\mu_1, ..., \mu_m\}$. Each generated PELC will be defined either according to the final target or to the obstacle$_{id}$ (cf. Algorithm 4). It is to be noted that if $m = 1$, the idea is to generate an optimal PELC* (cf. section 4.3.1) in clockwise and another in counter-clockwise direction respectively. In addition, if $m > 1$, then the chosen fixed values of μ are those which show the large shape possibilities of the PELC$_i^j$ (slow and quick convergence toward the PEI$_{id}$ (cf. Figure 2.5(a), page 34), where $v_j \otimes \Re_{id}$). For example if the slow and quick convergence correspond respectively to μ_{min} and μ_{max}, then the m values of $\mu \in S_\mu$ will be $\{\mu_{min}, \mu_{min} + \delta\mu, ..., \mu_{min} + (m-1)\delta\mu, \mu_{max}\}$, where $\delta\mu = (\mu_{max} - \mu_{min})/m$. Obviously, more is important the value of m closest is gPELC* to the optimal effective path (linking P_0 to P_f). It is to be noted that if T has $n+1$ vertices, thus T has n edges. The tree's size, given by $|E|$, corresponds to the number of edges. The number of vertices if each vertex generates $2m$ children is given by the following formula:

$$1 + 2m + (2m)^2 + ... + (2m)^h = \frac{(2m)^{h+1} - 1}{2m - 1} \tag{4.4}$$

where h corresponds to the greatest level in T (called also the height of the rooted tree).

- A **valid global path**, defined by gPELC$_n$, is a path which starts from v_0 and reaches the vertex v_n (linked to the reference frame attributed to the main target $\Re_f$, it is noted therefore $v_n \otimes \Re_f$) without any obstacle collision. The gPELC$_n$ is obtained from an oriented graph given by a sequence of vertices $(v_0, v_1, \ldots, v_n) \in V^n$ such that v_{i-1} is adjacent to v_i and $i \in \{0, 1, ..., n\}$. gPELC$_n$ is a path of length n from v_0 to v_n. It is to be noted that the indexes given to v_i are variables and are not related to any canonical labeling of the vertices but only to their position in the sequence.

- The optimal path gPELC* is a **valid global path** that over all possible gPELC$_n$

minimizes the function:

$$G = \sum_{i=1}^{n} J_{i-1}^{i}$$
$$= w_1 G_1 + (1 - w_1) G_{1Bis} + w_2 G_2 + w_3 G_3 + \tag{4.5}$$
$$w_4 G_4 + w_5 J_{\text{Distance\_gPELC\_FinalTarget}}$$

where:

- $w_i \mid i = 1..4 \in \mathbb{R}^+$ are the constants defined in equation 4.1, which permit the right balance between the different sub-criteria characterizing each elementary computed PELC, to give form to a gPELC. $w_5 \in \mathbb{R}^+$ permits us to give more interest to the gPELC_n which has a closest final point ($\text{gPELC}_n(s_f)$ where s_f, the final curvilinear abscissa of gPELC_n), to the final target P_f. This last information is embedded in the sub-criteria $J_{\text{Distance\_gPELC\_FinalTarget}}$.

- The global criteria G is defined therefore according to the vector $W_{gPELC} = \{w_1, w_2, w_3, w_4, w_5\}$ and the sum of the elementary sub-criteria (cf. section 4.3.1):

  ○ $G_1 = \sum_{i=1}^{n} J_{(i-1)\text{DistanceToPTi}}^{i}$

  ○ $G_{1Bis} = \sum_{i=1}^{n} J_{(i-1)\text{PELC}_{\text{Length}}}^{i}$

  ○ $G_2 = \sum_{i=1}^{n} J_{(i-1)\text{PELC}_{\text{Curvature}}}^{i}$

  ○ $G_3 = \sum_{i=1}^{n} J_{(i-1)Bool_{\text{MaximumCurvature}}}^{i}$

  ○ $G_4 = \sum_{i=1}^{n} J_{(i-1)Bool_{\text{Collision}}}^{i}$

To summarize, the idea to obtain the optimal gPELC* (cf. Algorithm 4) is to get the optimal sequence of elementary PELC to reach the main target P_f. The proposed Algorithm 4 permits us to obtain a tree T, containing vertices linked with PELC without collisions, and with each edge weight obtained while using equation 4.1. Each valid gPELC enables us to start from the vertex v_0 to reach vertices $\infty \mathfrak{R}_f$. The gPELC* is the one which minimizes G (cf. equation 4.5). Finally, the gPELC* contains the optimal sequence of local PELC_i^j (with their values r_j and μ_j). Generally, once the tree T is available, the optimal-path from the root v_0 to a vertex v_n ($v_n \infty \mathfrak{R}_f$) can be obtained while using for instance, tree-search-based Dijkstra's algorithm [Dijkstra, 1959] or the well-known Bellman–Ford algorithm.

It is to be noted that the **Else block** given between line 13 and 28 of Algorithm 4, expresses the fact that, when an obstacle$_{id}$ obstructs one extended PELC_i^f then this obstacle$_{id}$ will be selected as an intermediate orbit (before reaching after the Y-axis of $\mathfrak{R}_f$). This will be done while computing another PELC_i^{id}, starting from the same vertex v_i but aiming to terminate in the Y-axis of $\mathfrak{R}_{id}$ before adding a new vertex to T, enabling us to explore further this new branch. Nevertheless, if this new computed PELC_i^{id} collides with any other obstacle (before reaching the Y-axis of $\mathfrak{R}_{id}$), this branch is terminated without adding any new vertex to T. This choice

---

Algorithm 4: Overall proposed methodology to obtain gPELC*

Data: Initial vehicle state; Environment features: Obstacles$_{id=1,..,N}$ and Final target position;
$S_\mu = \{\mu_{k=1,...,m}\}$ the set of possible μ values.
Result: gPELC* (optimal global Parallel Elliptic Limit-Cycle)

1 //Initialization
2 $\mathcal{S}_0 = \{v_0\}$; //The set of the vertices at **Level**$_0$ of the tree T//
3 $i = 0$;
4 **while** <u>Not all the vertices of $\mathcal{S}_i \infty \mathfrak{R}_f$</u> **do**
5     $i = i + 1$;
6   //Compute all the PELC: starting from vertices given in $\mathcal{S}_{i-1}$ (which are not ∞ with $\mathfrak{R}_f$) and targeting $\mathfrak{R}_f$//

7     $\mathrm{PELC}^f_{[j=1,...\mathrm{Card}(\mathcal{S}_{i-1})]}(\mathrm{PEI}_f, r = \pm 1, S_\mu)$;

8   **forall** <u>Obtained PELC^f_j</u> **do**
9   **if** <u>It does not collide with any obstacle</u> **then**
10   //$v_{f\#}$ exists thus, where "$_{f\#}$" corresponds to the index number of the vertex ∞ $\mathfrak{R}_f$. This implies that a valid global path could be obtained while knowing all the antecedents vertices of $v_{f\#}$//
11            $\mathcal{S}_i \leftarrow v_{f\#}$; //Add the vertex to the tree T//
12            $e^{f\#}_j \leftarrow J(\mathrm{PELC}^f_j(\mathrm{PEI}_f, r, \mu_k))$; //Compute the weight of the edge $e^{f\#}_j$//
13   **else**
14   Obtain the id of the first obstructing obstacle;
15   //Compute all the PELC: starting from the vertex v_j and finishing in $\mathfrak{R}_{id}$//
16            $\mathrm{PELC}^{id_{r\mu}}_j(\mathrm{PEI}_{id}, r = \pm 1, S_\mu)$; //Where the index number "$_{id_{r\mu}}$", linked to the value of r and μ, corresponds to the vertex index ∞ $\mathfrak{R}_{id}$//
17   **if** <u>It does not collide with any other obstacle</u> **then**
18   **if** <u>Exhaustive Expanding Tree</u> **then**
19   //Add all valid vertices
20                    $\mathcal{S}_i \leftarrow v_{id_{r\mu}}$;
21                    $e^{id_{r\mu}}_j \leftarrow J(\mathrm{PELC}^{id_{r\mu}}_j(\mathrm{PEI}_{id}, r, \mu_k))$;
22   **else**
23   //Add only the optimal vertex in each direction (clockwise and counter-clock.)
24                    $\mathcal{S}_i \leftarrow v^*$;
25                    $e_j \leftarrow J^*$;
26   **end**
27   **end**
28   **end**
29   **end**
30 **end**
31 //Apply Dijkstra's algorithm [Dijkstra, 1959] on the obtained final tree T//
32 gPELC* = DijkstraAlgorithm(T);

---

is made to avoid infinite loops and to reduce the number of combination given by Algorithm 4. This supposition was also made because if another collision is taken into account, it is obligatory to take this new obstacle id as a new intermediate orbit, which is in contradiction with the first supposition, consisting of addressing the case of obstacle$_{id}$ only because it is the first obstructing obstacle which does not allow PELC^f_i to reach the Y-axis of the main target $\mathfrak{R}_f$ (from the vertex v_i).

Inside this same **Else block** (line 13 to 28) there is also an important characteristic to highlight. It corresponds to the **If block** (between line 17 and 27) which permits us to apply either an Exhaustive Expanding Tree (EET) or not. If yes, this corresponds to adding to the tree all the valid vertices and if no, then adding only the optimal vertices. Consequently, in the first case, the number of branches (PELC) from the vertex $v_{i-1} \otimes \mathfrak{R}_{i-1}$ to the reference frame $\otimes \mathfrak{R}_i$ could be at maximum equal to $2m$ (if all the computed PELC are valid) and in the second case, the maximum number of branches from v_{i-1} is 2 (which correspond respectively to the optimal PELC* for clockwise and counter-clockwise directions). The main objective of the second case is to reduce the number of possible expanding branches at each iteration of Algorithm 4. The aim is obviously to reduce the overall computation time to obtain the gPELC*. In this last version, the number of explored states is reduced but does not guarantee to have an even a good solution gPELC* as given by EET. This result will be highlighted in the simulations given in section 4.4.4. In these simulations, it is shown also that the solutions obtained with only optimal vertices expanding are generally not so far from the effective gPELC* (given by EET).

In what follows, a Hybrid$_{RC}$ and Homogeneous Control Architecture (HHCA) will be detailed. This architecture uses notably the definition of optimal PELC* and gPELC* to obtain with homogeneous way the vehicle's set-points.

4.4 HOMOGENEOUS AND HYBRID$_{RC}$ CONTROL ARCHITECTURE

The proposed Hybrid$_{RC}$ and Homogeneous[3] Control Architecture (HHCA, cf. Figure 4.2) aims to simplify, manage and control either reactive or cognitive vehicle navigation (cf. section 1.3.2, page 11). By reactive, it is mainly meant that the navigation of the vehicle is done with the minimum information on the environment, whereas cognitive control is based on quasi-full knowledge on the environment. Obviously, when this knowledge is available, the cognitive control permits us to lead generally to optimal (or sub-optimal) robot navigation. Nevertheless, it needs a lot of processing time to reach this solution which could lead, in certain cases, to an unusable approach (cf. section 1.3.2, page 11).

Blocks numbered 1 to 3 in Figure 4.2 are in charge of detecting/localizing/characterizing any important features in the environment. Mainly, these blocks must provide the list of all perceived obstacles (or known according, for instance, to a road map) and the target to reach. Any possible obstructing object (obstacle/wall/pedestrian/ etc.) is characterized as specified in section 2.1 (page 26) by an ellipse given by the parameters (h, k, Ω, A, B). This characterization can be ensured even off-line (using for instance a road map of static objects) or on-line using for instance a camera positioned in the environment [Benzerrouk et al., 2009] or the robot's infrared sensors [Vilca et al., 2012c] (cf. section 2.5.2, page 40).

This architecture permits us to simply link the set-points defined by the cognitive or reactive levels to the effective achievement of the vehicle's movements. It is to be noted that the proposed HHCA uses the flexible and safe PELC (cf. section 2.3.1, page 31) to perform either reactive (cf. section 4.4.1) or cognitive (cf. section 4.4.2) vehicle navigation. Indeed, the PELC is used here either as a local planner (instan-

---

[3]Homogeneous in terms of used set-points and control law.

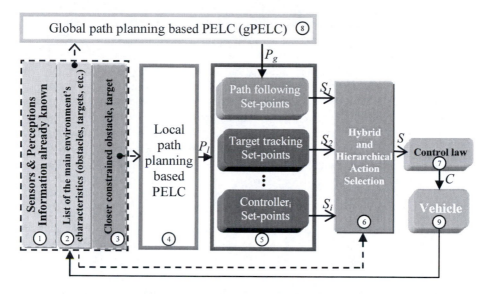

FIGURE 4.2 (See color insert) The proposed Homogeneous and Hybrid$_{RC}$ Control Architecture (HHCA) for mobile robot navigation.

taneous or short-term) or as a global planner (more cognitive approach while using gPELC (cf. section 4.3.2)) when the overall environment is well known. It is important to mention also since the HHCA (cf. Figure 4.2) uses the uniform set-points convention (cf. section 2.4, page 37) for performing either cognitive or reactive mode. These set-points are taken within the PELC-generated trajectories (cf. section 4.3). Indeed, the proposed HHCA architecture contains different set-point blocks which have as input the PELC already defined either locally (block number 4 in Figure 4.2 for reactive navigation) or globally (block number 8 in figure 4.2 for cognitive navigation). Once homogeneous set-points are obtained (block 5) according to what is defined in section 2.4(page 37), the common control law defined in section 3.2.2 will be used to stabilize the error to zero (cf. Figure 4.2).

Through the overall proposed control architecture, the objective is to highlight its genericity, flexibility and its reliability to deal with a large variety of environments (e.g., cluttered or not; dynamic or not, etc.). This architecture could adapt itself according to the navigation context and the occurrence of new events. It is shown in what follows:

- the way to act on-line to a changing environment when the used navigation strategy is reactive (cf. section 4.4.1),

- the used cognitive navigation (cf. section 4.4.2).

Otherwise, the choice between reactive and cognitive mode, according to the navigation context, will be illustrated through the proposed hybrid and hierarchical action selection process (cf. section 4.4.3). A multitude of simulations are given in section 4.4.4 to highlight the potentialities of the proposed overall control architecture.

4.4.1 Reactive navigation strategy based on local PELC*

This section deals with the challenging issue of on-line mobile robot navigation in unknown and cluttered environments. Indeed, it is considered in what follows, a mobile robot discovering the environment during its navigation; it should react to unexpected events (e.g., obstacles to avoid) while guaranteeing to reach its objective. In fact, reactive navigation is most desirable when the environment is not well known or in the case where this environment is highly dynamic and/or uncertain [Adouane, 2009b]. It is shown more specifically in what follows the use of PELC* for reactive navigation. The robot aims to navigate from its initial position to the final target while using only local defined PELC* and while dealing sequentially with the obstructing obstacle/wall/etc. This reactive navigation supposes nevertheless the efficiency and the reliability to obtain on-line the features of the obstacles (cf. section 2.5.2, page 40).

The proposed Algorithm 5 activates the obstacle avoidance behavior as soon as there exists at least one object which can obstruct the future robot movement toward its target. Otherwise, the behavior of target reaching (still while using PELC*) is activated. It is mentioned in Algorithm 5, the notion of "Current final target" because the robot is supposed to have a multitude of sequential targets to reach (cf. chapter 5). The good performance of the reactive navigation needs to manage some conflicting situations which could, in certain cases, lead to trajectory oscillations or dead-ends. Several reactive rules are detailed in [Adouane, 2009b] to avoid these situations. For instance, it has been proposed to maintain the direction of avoidance (clockwise or counter-clockwise) when the robot avoids two consecutive obstacles (without finishing yet the avoidance of the first, therefore the robot does not yet reach the first obstacle axis Y_{OT} (cf. Figure 2.5(a), page 34)).

The main differences between the current presented reactive navigation (cf. Algorithm 5) and the fully reactive navigation given in section 2.5 (page 39) are summarized below:

- The PELC set-points are used here for both obstacle avoidance and for target reaching controllers. In the previous architecture, the target reaching controller uses another set-point definition (cf. section 2.5.4.2, page 47).

- When the obstacle avoidance is active in fully reactive navigation, the direction (clockwise or counter-clockwise) and the robot's behavior (Attraction or Repulsion) are determined exclusively according to simple and deterministic rules, and the value of μ is also fixed once and for all. In the current less reactive navigation all the above choices are made by optimization of the criteria J (cf. equation 4.1).

Algorithm 5: Reactive navigation using optimal PELC* paths

Input: All the features h, k, Ω, A, B of the closest constrained obstacle (cf. Algorithm 2); Value of K_p (the desired minimum safe distance "offset" to the obstacles); Current final target localization (x_f, y_f).

Output: Current PELC*$((h, k), \Omega, (A, B, K_p), \mu^*)$ to follow (cf. equation 2.3 (page 32) and section 4.3.1).

1 **if** It exists at least one obstructing obstacle (cf. Algorithm 2, page 43) **then**

2   //Obstacle avoidance behavior

3   Obtain the direction r and μ^* optimizing the PELC according to the criteria J (cf. section 4.3.1).

4   PELC*$((h, k), \Omega, (A, B, K_p), \mu^*)$;

5 **else**

6   //Target reaching behavior

7   PELC*$((x_f, y_f), 0, (\xi, \xi, \xi), \mu^*)$;

8   Where ξ is a positive very small value $\longrightarrow 0$

9 **end**

4.4.2   Cognitive navigation based on gPELC*

Reactive navigation as given above could be optimal to avoid a single obstacle but not optimal at all if the navigation strategy needs to take into account several obstacles before reaching the final target. To perform cognitive navigation, this implies obviously much more knowledge about the environment (generally all the free and obstructed spaces) than for reactive navigation. The robot must define an overall path/trajectory/waypoints while taking into account the possible multiple obstacles in the environment.

To perform cognitive navigation, the proposed HHCA uses the overall optimized paths obtained according gPELC* (cf. section 4.3.2). Several simulations will be shown in section 4.4.4.

4.4.3   Hybrid$_{RC}$ and hierarchical action selection

As given in section 1.4 (page 15), there are two coordination processes to manage the activity of multi-controller architectures. The HHCA (cf. Figure 4.2) is based on the *action selection* process. It is called the Hybrid and Hierarchical Action Selection process and is summarized in Algorithm 6. This process aims to activate either reactive or cognitive navigation according to the environment knowledge and perceptions.

The cognitive navigation is activated only if the entire environment is well known or when the navigation is achieved in relatively low dynamic environment; low enough so that the gPELC* (cf. section 4.4.2) could be re-computed on-line. In the case that it cannot be obtained on-line, instead of stopping the vehicle's navigation

---

Algorithm 6: Hybrid and hierarchical action selection process

Data: Environment knowledge and perceptive information
Result: The more appropriate navigation strategy

1 **while** Final target is not reached **do**
2   **if** gEPLC* exists **then**
3   //Cognitive navigation
4   //gEPLC* exists means that the overall environment
5   //knowledge is available
6   Path following control activation w.r.t. gPELC*;
7   **else**
8   //Reactive navigation
9   //Defined w.r.t. the current obtained PELC*
10   **if** Static obstacles & Certain environment **then**
11   Local path following control activation;
12   **else**
13   Target reaching control activation;
14   **end**
15   **end**
16 **end**

(which could be an option), the vehicle will switch to navigate in a reactive way. This last navigation could be done in two ways: the first consists of using path following control, based on local computed PELC*, in the case where the current obstructing obstacle is static and could be accurately detected; the second reactive navigation is performed if the environment is dynamic and/or with a lot of uncertainty, in this case the vehicle has to navigate with even more reactivity (no pre-planned path to follow), using on-line target tracking control (cf. section 2.4.2, page 38). The value of the R_S radius (cf. Figure 2.6(b), page 37) could be fixed according to the measured uncertainty rate and to the dynamicity of the obstacles. This implies obviously to have specific metrics to characterize the environment uncertainty and the dynamicity of the detected obstacles (velocities, acceleration, etc.). These two criteria are more linked to the perceptive aspects and are still an open and active research area.

4.4.4 Extensive validation by simulation

Let us show in what follows several examples to exhibit the flexibility and efficiency of the proposed HHCA. The simulations were implemented using MATLAB® software (the porting to C++ language will be done in the near future to enhance the processing time) and performed with an Intel Core I7, CPU of 2.70 GHZ and a RAM of 32 GO. The robot kinematics is based on a tricycle model (cf. equation 3.8 (page 58) and Figure 3.3 (page 58)); its features are:

- $l_b = 12$ cm and $\gamma_{max} = 45°$.

- To characterize the robot collisions with the environment, the robot is surrounded completely by an ellipse (which has a major axis of 14 cm and a minor axis of 9 cm).

- The robot maximum field of view is considered as a circle centered on the robot with a radius of 72 cm (cf. Figure 4.7). This circle corresponds to the maximum distance where the robot can detect an obstacle. This perception is mainly used for safety behavior (automatic stop if the obstacle is too close) or in reactive navigation.

- The control law parameters are given by $\mathbf{K} = (10, 5, 2, 0.3, 5, 0.01)$ (cf. section 3.2.2, page 57).

The first set of simulations (cf. section 4.4.4.1) will show the use of the planned gPELC* for different kind of environments (cluttered, structured, etc.) and the other set of simulations (cf. section 4.4.4.2) will show the use of PELC and gPELC* to perform either reactive or cognitive navigation according to the environment state.

4.4.4.1 The use of gPELC* for different environments

4.4.4.1.1 **Cluttered environment**   Figure 4.3(a) shows the application of Algorithm 4 for a set of μ values $S_\mu = \{0.1, 0.4, 0.7\}$ in each direction (clockwise and counter-clockwise). Several simulations with their initial inputs and results are summarized in Table 4.1. This table shows, for instance, the weights $W_{gPELC} = \{w_1, w_2, w_3, w_4, w_5\}$ characterizing the global cost function (cf. equation 4.5) for each planned gPELC, the obtained tree characterizing the optimization process. In the simulation given in Figure 4.3(a), Algorithm 4 had to explore 31 vertices which required 9.48 s as computation time. This optimization permits us to obtain 14 valid gPELC and the optimal one (according to the applied global cost function) has an optimal cost of $G = 3.03$.

Figure 4.3(b) corresponds to a simulation which has the same initial inputs as in Figure 4.3(a) but while expanding the tree (given by Algorithm 4) for only the optimal vertices (cf. Table 4.1). According to this table, it is found that the computation time is reduced by 28%, and despite this, the obtained optimal gPELC* cost G is still very close to the obtained gPELC* given in Figure 4.3(a). The obtained gPELC* is nevertheless different in terms of path shape and intermediate cost function (as for G_2, where the obtained gPELC* is smother than that obtained in Figure 4.3(a) ($G_2 = 0.18$ instead of 0.25)). Figure 4.4(a) shows the obtained tree for the simulation given in Figure 4.3(b). Due to the smaller number of obtained vertices while using only optimal vertices expansion, this methodology will be preferred in the coming simulations to simplify and highlight better certain results. Figure 4.3(c) gives an example of the influence of the wights chosen for W_{gPELC} to obtain the optimal gPELC*. Indeed, while modifying these weights between simulations given in Figure 4.3(b) and Figure 4.3(c), the obtained final gPELC* has been changed (cf. Table 4.1).

Figure	Exhaustive Expanding Tree?	W_{gPELC}	Number of valid explored vertices	Number of valid gPELC	Time [s]	Optimal gPELC* cost function			
						G	G_1	G_{1Bis}	G_2
4.3(a)	Yes	{0.50, 0.0005, 10^6, 10^6, 10}	31	14	9.48	3.03	0.21	2.57	0.25
4.3(b)	No	{0.50, 0.0005, 10^6, 10^6, 10}	7	3	6.85	3.04	0.35	2.51	0.18
4.3(c)	No	{0.95, 0.0005, 10^6, 10^6, 10}	10	5	7.72	0.83	0.32	0.29	0.22
4.5(a)	No	{0.60, 0.0020, 10^6, 10^6, 10}	5	2	15.68	3.69	0.26	2.69	0.74
4.5(b)	No	{0.50, 0.020, 10^6, 10^6, 10}	9	2	6.19	11.92	1.60	6.55	3.77
4.5(c)	No	{0.50, 0.020, 10^6, 10^6, 10}	14	4	6.36	11.40	1.02	6.65	3.73
4.4(b)	Yes	{0.60, 0.0020, 10^6, 10^6, 10}	49	18	82.12	4.91	0.58	3.45	0.88
4.4(c)	No	{0.60, 0.0020, 10^6, 10^6, 10}	21	5	49.63	5.00	0.64	3.42	0.94
4.6(a)	No	{0.60, 0.0020, 10^5, 10^6, 10}	13	5	28.07	3.46	0.42	2.33	0.71

TABLE 4.1 Optimization results to obtain optimal gPELC* for different environments according to Algorithm 4 parameters.

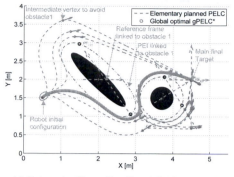

(a) Exhaustive Expanding Tree (cf. Algorithm 4)

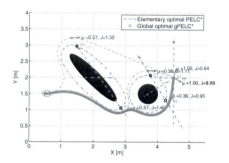

(b) Expanding Tree only for optimal node at each step

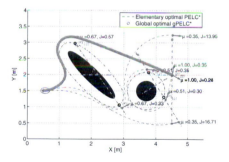

(c) Expanding Tree only for optimal node with other criteria parameters w.r.t. simualtion (b)

FIGURE 4.3   Global path planning based on gPELC*.

While simulations given in Figures 4.3(a) to 4.3(c) show the application of Algorithm 4 for a quite simple environment (2 obstacles), the simulations given in Figures 4.4(b) and 4.4(c) permit, among other things, to highlight the efficiency of the proposed Algorithm 4 for an even more cluttered environment (a configurations of 5 close obstacles) (cf. Table 4.1). These last simulations will be used in next subsec-

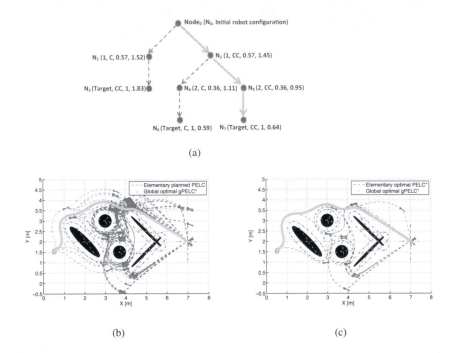

FIGURE 4.4 **(a)** Tree representation to obtain the gPELC* given in Figure 4.3(b). Each node $N_i|_{i=1..7}$ (except the root node N_0) is represented respectively by an index 1 or 2 (the $_{id}$ of the avoided obstacle) or target (for final target reaching); the direction of avoidance C or CC (for respectively clockwise or counter-clockwise); the value of μ^* and finally the value of the elementary obtained PELC* cost J^* (cf. equation 4.1). The green arrows correspond to the optimal solution **(b)** and **(c)** global path planning-based gPELC* using, respectively, Exhaustive Expanded Tree and only optimal ones (at each step).

tion 4.4.4.2, where the obtained optimal path gPELC* will be used as an initial path to show the flexibility of the overall proposed control architecture to switch easily and safely from cognitive navigation to a reactive one and *vice versa*.

4.4.4.1.2 Structured environment To emphasize the reliability of the proposed planning method-based PELC for different kinds of environments, it is also important, in addition to a cluttered environment, to verify its efficiency for a structured environment (as we can find in an indoor or urban environment with walls and right angles "perpendicular straight lines"). The obstacle modeling given in section 2.1 (page 26) is used in what follows (cf. Figure 4.5) to surround walls (or sidewalks for instance) of different dimensions. Each obstacle is surrounded with thin Surrounded Ellipse (given by SE_i in Figure 2.2) and with a Parallel Ellipse of Influence (PEI$_i$ in Figure 2.2 (page 27)) which will give a safe margin between the

robot and the obstacle. The first simulation given in Figure 4.5(a) shows the efficiency of Algorithm 4 even for a Trap configuration. The robot initial posture is
$(x_0, y_0, \theta_0, \gamma_0) = (0, 1.5, -45°, 0°)$. Algorithm 4 obtained an optimal gPELC* (cf.
Table 4.1) while avoiding locals minima [Ordonez et al., 2008]. According to Table
4.1, it is seen that the time necessary to obtain gPELC* is equal to 15.68 s which is
relatively high w.r.t. the other simulations which have only few valid explored vertices. This is explained by the fact that according to the configuration given by this
trap, a lot of iterations of Algorithm 4 lead to invalid vertices (thus, collision of the
computed PELC with an obstacle).

The simulations given in Figures 4.5(b) and 4.5(c) show a complex enough environment based on several walls/obstacles (forming labyrinthine corridors). The set
up difference between the simulation given in Figure 4.5(b) and the one given in
Figure 4.5(c) is in the value of the obstacle safe Margin (cf. section 2.3.1, page 31)
which corresponds to K_p (cf. Figure 2.2 (page 27) and equation 2.3 (page 32)). For
the first simulation (cf. Figure 4.5(b)) $K_p = 16$ cm whereas the second (cf. Figure
4.5(c)) $K_p = 32$ cm (cf. Table 4.1 for simulations parameters and optimization results). The second simulation restricts much more the possible robot states to reach
the final Target, but permits us to obtain much more safe gPELC* because the robot is
forced to navigate as far as possible from any obstacle. The obtained gPELC* is close
to one path which could be produced by a Voronoi method [Aurenhammer, 1991]
[Latombe, 1991], but with an important advantage in what we propose is that the
obtained path (gPELC*) takes into account:

- the kinematic and structural constraints of the robot (non-holonomy, maximum
  steering angle γ_{Max}, etc.), and

- multi-criteria optimization (cf. equation 4.5) that is not only linked to safety
  criteria as for Voronoi method.

It is to be noted that K_p attributed to each obstacle could be different between
them. This could be chosen according, for instance, to the obstacle's dynamic or to
its features (location, shape, etc.). It is to be noted also that in certain situations where
the PEI$_i$ attributed to obstacle$_i$ is in intersection with the PEI$_j$ of another obstacle$_j$
(as seen in Figure 4.5), it is important to introduce some simple rules to guarantee always the convergence of Algorithm 4. Indeed, when computing an elementary PELC
w.r.t. an obstacle$_i$, the axis Y of the reference frame linked to this obstacle (cf. section
2.3.2) could not be reachable without going inside PEI$_j$. This is due to the intersection of the PEI (attributed to the two or more obstacles). The condition of stopping
the avoidance w.r.t. an obstacle can never therefore be attained. Thus, if at least two
obstacles have an intersected PEI, the following rules, to end the computation of the
PELC$_i$, must be applied:

- End (stop and validate) the computed PELC$_i$, the first time that its x abscissa
  sign (w.r.t. the reference frame $\mathfrak{R}_i$ linked (∞) to obstacle$_i$) changes from "-" to
  "+".

- If the above rule is not yet verified and the current computed PELC$_i$ is going to be inside another PEI$_j$ (with $j \neq i$), therefore, stop the computation of PELC$_i$ at a predetermined distance DS before going inside PEI$_j$, a node is therefore added in Algorithm 4, enabling us to continue the expanding while avoiding a deadlock. For example, in Figure 4.5(b), the robot must avoid the obstacle$_2$ in a clockwise direction, but its PEI$_2$ intersects with PEI$_4$, the above rules were therefore applied.

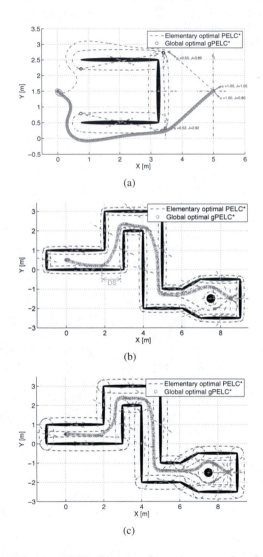

FIGURE 4.5 **(a)** Efficient gPELC* even for a trap configuration. **(b)** and **(c)** Global path planning-based gPELC* in structured environment, with respectively (b) $K_p = 16$ cm and (c) $K_p = 32$ cm.

4.4.4.2 Switch from cognitive to reactive navigation and vice versa

The aim of the following simulations is to show the high flexibility of the proposed control architecture to perform either reactive or cognitive navigation while guaranteeing smooth and steady robot behavior (cf. section 3.2.2, page 57). Thus, using a relatively simple algorithm (as proposed in Algorithm 6), the possibility of switching from cognitive to reactive mode and *vice versa* will be shown. It is to be noted that these simulations do not have as an objective to propose an optimal choice between the activation of one mode (cognitive or reactive) w.r.t. the other. This important issue is an open research area [Ranganathan and Koenig, 2003] [Ridao et al., 1999] [Rouff and Hinchey, 2011] [Mouad et al., 2012] and the inherent structure of the proposed control architecture is particularly appropriate to address this kind of interesting issue. This will be the subject of future investigations.

The robot navigation given in Figures 4.6(a) and 4.6(b) show the activation of different modes. The sequence given in Figure 4.6(a) is Cognitive → Dynamic obstacle avoidance and finally Cognitive navigation, whereas in Figure 4.6(b): Cognitive → Dynamic obstacle avoidance → Reactive navigation. In both simulations, the robot starts by performing path following (cf. section 2.4.1, page 37) of the already obtained gPELC* given in Figure 4.4(b) (cf. Table 4.1). Since the initial obtained gPELC* considers a static environment, the robot starts to follow this global optimal path to reach the final Target. At the instant 11 s, the obstacle$_3$ starts to move in a straight line (cf. Figures 4.7(a) to 4.7(c)). This movement makes unsafe the initial planned gPELC*, therefore, as soon as this obstacle is in the robot's field of view (represented by pink dashed circle in Figure 4.7), the robot starts to perform reactive obstacle avoidance using the local PELC* (cf. section 4.4.1). The used set-points in this navigation phase are based on target reaching set-points (cf. section 2.4.2, page 38) while taking the parameter $R_S = 0$ (cf. Figure 2.6(b) (page 37)).

Once obstacle$_3$ is completely avoided (cf. Algorithm 5), the robot continues to reach the final target by using either cognitive navigation (cf. Figure 4.6(a)) or reactive navigation (cf. Figure 4.6(b)). In the first case, a new gPELC* is recomputed (from the current initial robot's configuration) and followed by the robot. The replaned path features are given in the last row of Table 4.1. Figures 4.6(c) and 4.6(e) give the robot's navigation details of the simulation given in Figure 4.6(a), whereas Figures 4.6(d) and 4.6(f) give the navigation's details of Figure 4.6(b).

Concerning the simulations' indicators given in Figures 4.6(c) and 4.6(d), they show in general that in both simulations, the robot navigates far enough from the closest obstacle. Indeed, specifically in Figure 4.6(c), the distance evolution between the robot and the closest obstacle is always smooth and bigger than the fixed safe avoidance offset ($K_p = 22$ cm and represented in dotted red lines in the figures). There exists nevertheless, one critical phase in Figure 4.6(d) where the robot, in reactive mode, avoids obstacle$_5$ and starts to avoid obstacle$_4$. In this phase the distance robot–obstacle$_4$ is less than the offset but the robot remains far enough from the obstacle to avoid any collision (cf. Figure 4.6(b)). In fact, the offset value is fixed according first, to the dimension of the robot (surrounded by an ellipse with a major axis equal to 14 cm) and also to the robot's physical constraints and control reliabil-

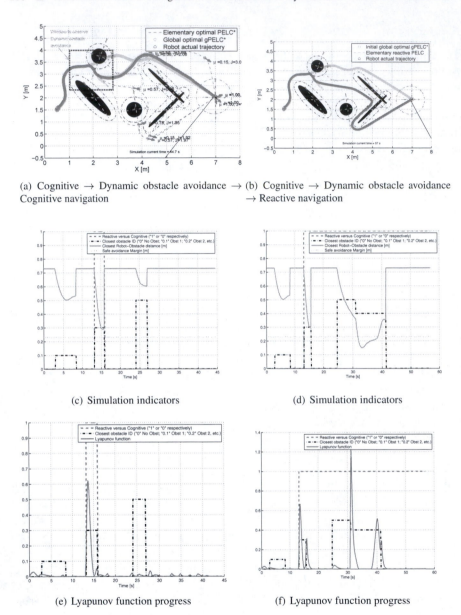

(a) Cognitive → Dynamic obstacle avoidance → (b) Cognitive → Dynamic obstacle avoidance
Cognitive navigation → Reactive navigation

(c) Simulation indicators   (d) Simulation indicators

(e) Lyapunov function progress   (f) Lyapunov function progress

FIGURE 4.6   (See color insert) Reactive versus cognitive navigation.

ity, etc. In the simulation given in Figure 4.6(d), the value of the minimum distance robot–obstacle$_4$ is equal to 15 cm, and therefore, in all cases, bigger than the major axis of the surrounded ellipse.

It is to be noted also that this critical situation happened because of the difficult initial robot configuration when obstacle$_4$ must be avoided. Indeed, when obstacle$_4$

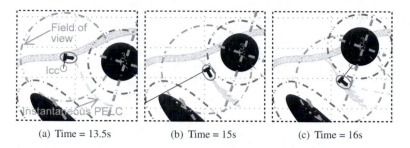

| (a) Time = 13.5s | (b) Time = 15s | (c) Time = 16s |

FIGURE 4.7 Dynamic obstacle avoidance. The cyan dashed lines in (a) to (c) correspond to the different instantaneous PELC* computed by HHCA when the robot avoids Obstacle$_3$. In (c) Obstacle$_3$ is completely avoided and a PELC* is computed w.r.t. the main target.

becomes the most obstructing obstacle (cf. Algorithm 2, page 43), the robot is very close to it and has an important angular error to the PELC set-point $\geq 90°$, therefore the robot is in a very difficult configuration to safely avoid obstacle$_4$. These specific critical situations are unfortunately unavoidable in reactive mode since the robot discovers its environment online [Adouane, 2009b] [Adouane et al., 2011]. An emergency stop could be obviously activated if the robot–obstacle distance is less than a certain value. In addition, as information, the computed PELC (to avoid obstacle$_4$) has a counter-clockwise direction to avoid obstacle$_4$. Thus, in the same direction then the avoidance of obstacle$_5$. Indeed, in reactive avoidance, if the robot starts to avoid an obstacle$_i$ and switches to another obstacle$_j$ (which has an intersection w.r.t. obstacle$_i$), the robot must follow the same direction (clockwise or counter-clockwise) as for the previous avoidance. This avoids the robot's dead-ends and infinite oscillations [Adouane, 2009b].

Figures 4.6(e) and 4.6(f) give the progress of the Lypapunov function (cf. equation 3.15, page 59). These figures show the stability of the system to make the errors converge always to 0, even when the robot enters reactive mode, where the obstacle to avoid can change suddenly according to the robot's perceptions (cf. Algorithm 2). In fact, when the robot starts to avoid another obstacle, another local PELC will be re-computed, taking into account the current obstacle features (position, orientation, dimension, etc.). This causes inevitably an abrupt jump in the error value (therefore on the Lyapunov function), but after that, this function decreases always until reaching 0, which attests to the asymptotic stability of the overall control architecture.

4.5 CONCLUSION

This chapter has focused on the aptitude of a control architecture to perform, either reactive or cognitive autonomous navigation, according to the vehicle/environment context. It has been presented therefore a Hybrid$_{RC}$ (Reactive/Cognitive) and Homogeneous (in term of set-points and control law) Control Architecture (HHCA) which uses mainly PELC trajectories as the main component for the different vehicle navigations. The main objective of this kind of Hybrid$_{RC}$ structure is to permit us to deal online and safely with unpredictable/uncertain situations, and also to optimize the overall vehicle navigation if the environment is well-known/mastered.

Whereas the previous chapters emphasized more reactive navigation, this chapter gave an overview of the different proposed planning methods and focused on the proposed techniques, based on multi-criteria optimization, for optimal short- and long-term path planning. More specifically, for short-term planning it has been presented optimal PELC (PELC*) which is obtained while including several sub-criteria and constraints, among them: the robot's initial state and structural constraints (non-holonomy and maximum steering); the enhancement of smoothness, safety of the obtained trajectories as well as the minimization of the robot's traveled distance. Furthermore, to perform appropriate cognitive navigation, it is important to have a long-term planning technique. Hence, it has been proposed to appropriately sequence a multitude of PELCs to obtain optimal global path planning-based PELC (gPELC*) (cf. Algorithm 4).

The HHCA has been designed in order to use homogeneous set-points-based PELC* or gPELC*. It is important to mention also that either in reactive or cognitive navigation, the vehicle is controlled with the same control law. The overall HHCA stability (based on Lyapunov definition) could be therefore rigorously demonstrated and analyzed (cf. chapter 3). Otherwise, an appropriate hybrid and hierarchical process of selection has been designed to manage the different navigation contexts and sub-tasks.

Through the overall proposed multi-control architecture, with its different blocks and mechanisms, the objective is to highlight its genericity, flexibility and reliability to deal with a large variety of environments (cluttered or not, structured or not, and dynamic or not) while guaranteeing the smoothness of the switch between the different vehicle's navigation modes. An extensive number of simulations, with several situations, have been performed to confirm the potentialities of the proposed HHCA.

Obviously, the HHCA structure does not permit us to resolve all the aspects linked to the hybridization (reactive / cognitive), but at least it permits us to give an overall structure / mechanism to clarify and isolate the main components which still need important developments. The formulation of the optimal balance between reactive and cognitive navigation is among the most important issues to resolve in the near future. This will be resolved notably by using appropriate metrics to better characterize the environment (dynamicity / uncertainty, etc.).

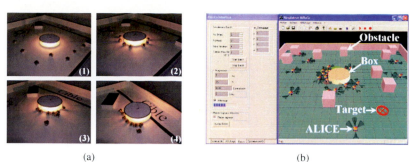

(a)   (b)

COLOR FIGURE 1   (a) Experimentation of the proposed control architecture with 8 mini-robots ALICE pushing a cylindrical object to a final assigned area; (b) *MiRoCo* simulator.

(a)   (b)

(c)   (d)

(e)   (f)

COLOR FIGURE 1.1   Different kinds of mobile robot locomotion. (a) AQUA® bio-inspired AUV (KROY version) [Speers and Jenkin, 2013], (b) Piranha® USV from Zyvex® company, (c) UAV from Fly-n-Sense® company, (d) two LS3s from Boston Dynamics®, (e) the first two-wheel self-balancing chair Genny®, inspired by Segway®, (f) Robot Curiosity (from NASA) exploring Mars planet.

(a) 1949 - Turtle-like robot of W. Grey Walter

(b) 1967 - Shakey mobile robot

(c) 1987 - VaMoRs Autonomous Van

COLOR FIGURE 1.2   UGV's short historic aspect.

(a) Google® Car

(b) SARTRE project

(c) GCDC project

(d) VIPALAB® vehicles

COLOR FIGURE 1.3   Different autonomous UGVs projects in several environments.

(a) KIVA® system

(b) Autonomous UGV de- (c) Autonomous shuttles
polyed by TEREX® com- (from Ultra Global PRT) in
pany in ports Heathrow airport, United
Kingdom

COLOR FIGURE 1.6   Different actual deployed UGVs in closed environments.

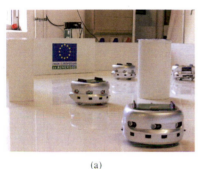

(a)

(b)

COLOR FIGURE 2.1   Autonomous navigation of (a) a group of mobile robots (Khepera®) and (b) an electric vehicle (Cycab®) in an urban environment (Clermont-Ferrand, France). MobiVIP project (Predit 3).

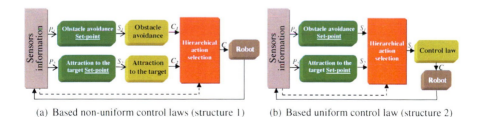

(a) Based non-uniform control laws (structure 1)   (b) Based uniform control law (structure 2)

COLOR FIGURE 2.7   The two main used multi-controller structures.

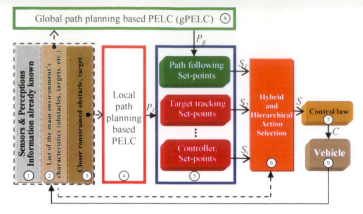

COLOR FIGURE 4.2 The proposed Homogeneous and Hybrid$_{RC}$ Control Architecture (HHCA) for mobile robot navigation.

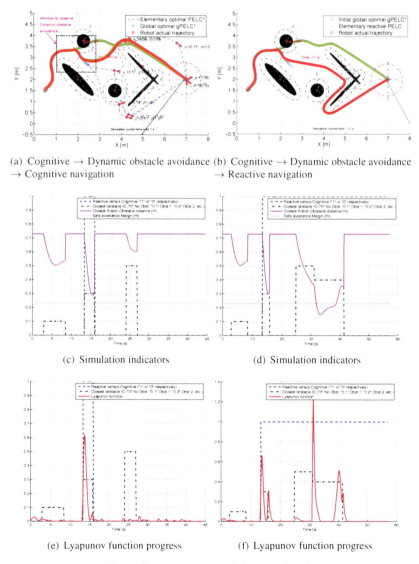

(a) Cognitive → Dynamic obstacle avoidance → Cognitive navigation

(b) Cognitive → Dynamic obstacle avoidance → Reactive navigation

(c) Simulation indicators

(d) Simulation indicators

(e) Lyapunov function progress

(f) Lyapunov function progress

COLOR FIGURE 4.6 Reactive versus cognitive navigation.

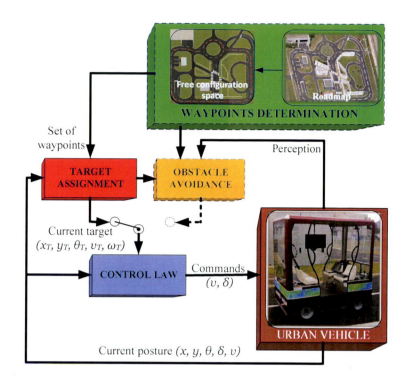

COLOR FIGURE 5.3 Proposed multi-controller architecture to perform autonomous vehicle navigation based on sequential target reaching.

COLOR FIGURE 5.27 Some images from the performed experiment.

(a) Modular robotics and self-asembling tasks (b) Swarm robotics and heterogeneous Robot-Robot cooperation (c) Robot-Robot and Robot-Human cooperation

COLOR FIGURE 6.1 Different cooperative robotics projects/tasks. (a) M-TRAN project [Murata and Kurokawa, 2012], corresponds to modular robots which can autonomously reconfigure themselves to form different 2D or 3D structures. (b) Swarmanoid project; Cooperation of a swarm of ground and aerial robots to achieve complex task [Mathews et al., 2015]. (c) Object handling by multiple mobile manipulators in cooperation with a human [Hirata et al., 2007].

(a) From left to right: proposed sequence strategy to lift any payload (shape/weight) on the robots.

(b) (c)

COLOR FIGURE 6.2 Cooperative manipulation and transportation of any payload shape/weight. (a) and (b) Multi-body dynamic simulation (with ADAMS®) for payload lifting using the developed strategy of positioning and lifting. (c) First prototype using Khepera® mobile robots and designed end-effector.

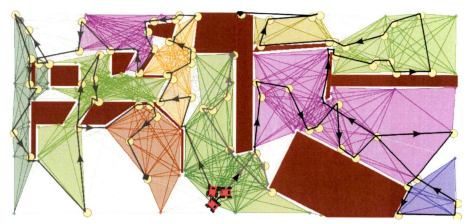

(a) Task allocation for 3 mobile robots while using appropriate waypoints definition and decentarlized exploartion policies [Lozenguez et al., 2012b].

(b)   (c)

COLOR FIGURE 6.3   Cooperative coordination and navigation using a group of Pioneer® in PAVIN (cf. Annex A, page 185) for (b) free area or (c) urban area [Lozenguez, 2012].

COLOR FIGURE 6.4 Example of autonomous navigation in formation of a group of UGVs in an urban environment (Clermont-Ferrand, France). Mo-biVIP project (Predit 3).

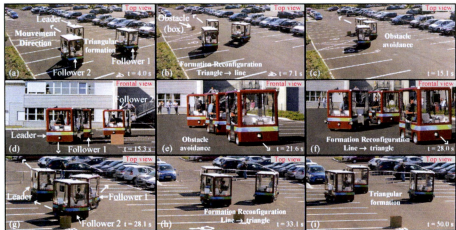

(a)

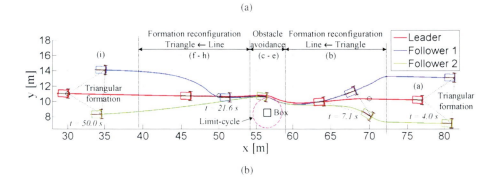

(b)

COLOR FIGURE 6.24 Final demonstration given in the context of the Safe-Platoon project.

CHAPTER 5

Flexible and reliable autonomous vehicle navigation using optimal waypoints configuration

CONTENTS

T HIS CHAPTER emphasizes the fact that it is not absolutely mandatory (as commonly admitted and broadly used in the literature) to have a predetermined trajectory to be followed by a robot to perform reliable and safe navigation in an urban and/or cluttered environment. A new definition of the navigation task, using only discrete waypoints in the environment, will be presented and applied for an urban electric vehicle. This approach permits us to reduce the computational costs and leads to an even more flexible navigation with respect to traditional approaches (mainly if the environment is cluttered and/or dynamic). In addition, several techniques are presented in this chapter to obtain the appropriate set of waypoints to perform reliable navigation; the most important is called OMWS-ET (for Optimal Multi-criteria Waypoint Selection based on Expanding Tree).

5.1 MOTIVATIONS AND PROBLEM STATEMENT

5.1.1 Motivations

As given in section 1.5 (page 21) the most popular works in literature use generally a pre-determined trajectory as navigation set-points. Section 1.5 details the most important characteristics linked to this kind of navigation. Among the most important, which could add complexity to performing this kind of navigation, let us cite for instance the need for specific planning method to generate the trajectory, the need to guarantee the continuity between different path segments, and the complexity of the replanning phase, etc.

 This chapter assumes that to perform safe and flexible navigation of a vehicle it is not mandatory to have a specific pre-planned reference trajectory. It presents the idea to use only a set of waypoints, appropriately disposed in the environment, to perform such navigation. The use of only a discrete number of waypoints in the environment will permit even more flexibility of the vehicle's movements, since it is allowed to perform more maneuvers between waypoints, while remaining obviously safe (non-collision of the vehicle w.r.t. the road limits or any obstructing obstacle). Hence, navigation using only waypoints allows us to avoid any path/trajectory planning which could be time-consuming and complex, mainly in cluttered and dynamic environments. Moreover, this kind of navigation does not require knowledge of the pose of the closest point to the followed trajectory (w.r.t. the robot configuration) and/or the value of the curvature at this point [Gu and Dolan, 2012]. Consequently, the navigation problem is simplified to a waypoint-reaching problem, i.e., the vehicle is guided by waypoints instead of following a specific fixed path (cf. section 1.5, page 21).

Moreover, it is important to notice that if the successive waypoints are closer to each other, then the vehicle tends to perform a path-following navigation. The proposed technique tends therefore to gather the different navigation techniques. In addition, the use of only waypoints to control the vehicle instead of a fixed trajectory, allows the robot to carry out local operations (to avoid such obstacle) while maintaining overall stability of the used hybrid multi-controller architectures (cf. section 5.2.1). This chapter particularly focuses on the problem of autonomous navigation of vehicles in an urban environment (cf. Figure 2.1(b), page 26). Several simulations and experiments, using a single or a group of VipaLab (multi-vehicle navigation) were performed showing the flexibility, reliability and efficiency of the developed navigation strategy (cf. section 5.5).

5.1.2 Problem statement

An important challenge in the field of autonomous robotics consists of ensuring safe and flexible navigation in a structured environment (cf. Figure2.1(b) and 5.1). In this work, safe navigation consists of not crossing over the road limits and bumping into obstacles while respecting the physical constraints of the robot. Flexible navigation consists of allowing different possible movements to achieve the task, while guaranteeing a smooth trajectory of the robot. The main idea of the proposed work is to guarantee both criteria simultaneously. The following scenario is considered (cf. Figure 5.1):

- The structured environment is a known road map where the roads have a specific width w_R.

- The robot model (kinematic) is known.

- The robot starts at the initial pose P_i and it has to reach the final pose P_f (in certain conditions, $P_i = P_f$).

As presented in section 1.5 (page 21) and according to the presented scenario, a safe reference path in a static environment can be obtained by different algorithms such as a Voronoï diagram [Latombe, 1991], potential fields [Khatib, 1986] or others [LaValle, 2006]. In the presented case, specific key positions should be defined in the static environment, and are called *waypoints*. Their numbers and configurations in the environment are detailed in section 5.4. Consequently, the navigation problem is simplified to a waypoint-reaching problem, i.e., the robot is guided by the waypoints (cf. Figure 5.2) instead of following a specific fixed path. The robot has thus to reach each waypoint with a defined position, orientation and velocity while satisfying distance and orientation error limits (E_{dis} and E_{angle} respectively) to perform safe navigation (cf. subsection 5.3.1).

This chapter is organized as follows: Section 5.2 presents the proposed navigation strategy based on sequential target reaching. In section 5.3, the control aspects will be detailed (architecture, stability/reliability, smoothness, etc.). Section 5.4 is dedicated

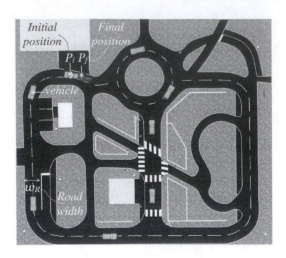

FIGURE 5.1   Nominal scenario with a road map and the task to achieve by the robot in its environment.

to an overview of the different proposed techniques to obtain the most appropriate set of waypoints. A large number of simulations (cf. sections 5.3.3 and 5.4.4) and experiments (cf. section 5.5) demonstrate the reliability of the proposed strategy of navigation. Finally, section 5.6 provides a conclusion for this chapter.

5.2   STRATEGY OF NAVIGATION BASED ON SEQUENTIAL TARGET REACHING

The proposed navigation strategy uses a sequence of N sorted waypoints appropriately disposed in the environment. The aim of this sequence is to guarantee safe and flexible robot navigation. Each waypoint $T_j = (x_{T_j}, y_{T_j}, \theta_{T_j}, v_{T_j})$ corresponds to a specific key configuration in the environment (cf. Figure 5.2). T_j is characterized by:

- A position (x_{T_j}, y_{T_j}).

- An orientation θ_{T_j} such as:

$$\theta_{T_j} = \arctan\left((y_{T_{j+1}} - y_{T_j})/(x_{T_{j+1}} - x_{T_j})\right) \tag{5.1}$$

  where: $(x_{T_{j+1}}, y_{T_{j+1}})$ corresponds to the position of the next target T_{j+1}. T_j orientation is therefore always oriented toward the waypoint T_{j+1}.

- A velocity v_{T_j}. It is important to mention that to perform the proposed navigation by reaching sequential waypoints (targets), it is mandatory to reach each target (except the final waypoint) with a velocity $v_{T_j} \neq 0$ to not have a jerky vehicle movement, at the starting and the arrival phase for each waypoint. The

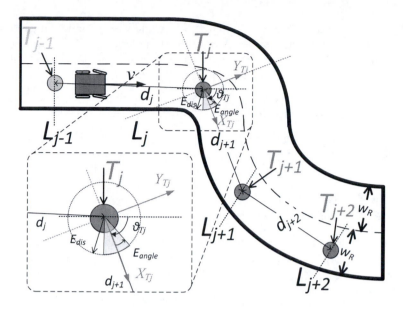

FIGURE 5.2   Description of waypoints assignment.

overall vehicle navigation becomes therefore smoother without oscillations in terms of linear velocity.

Different methods to obtain the appropriate set of waypoints (target set-points $((x_{T_j}, y_{T_j}, \theta_{T_j}, v_{T_j})\,|_{j=1...N}))$ are presented in section 5.4. They are based either on the heuristic method or on multi-criteria optimization.

To define the robot's navigation strategy between the successive waypoints (cf. subsection 5.2.2), an orthogonal reference frame $X_{T_j} Y_{T_j}$ (cf. Figure 5.2) is attributed to each waypoint, where:

- the X_{T_j} axis connects the position of T_j to the following waypoint T_{j+1} and oriented toward T_{j+1}, and

- the Y_{T_j} axis is perpendicular to X_{T_j} and is oriented while following trigonometric convention.

This reference frame will be used in subsection 5.2.2 to perform the target assignment process. In addition, to insure safe robot navigation between successive waypoints, each waypoint is assigned upper error bounds defined by E_{dis} and E_{angle} (cf. Figure 5.2). They correspond respectively to the maximal distance d and angle e_θ errors between the robot and the target (cf. Figure 3.3, page 58) when it crosses the axis Y_{T_j}. Further, E_{dis} and E_{angle} correspond to a kind of maximal error tolerance when the robot reaches the target T_j. This tolerance is notably related to the inaccuracies of the robot localization and/or to the performance of the used control law. The maximum authorized values of E_{dis} and E_{angle} allow us to keep reliable robot navigation toward the target T_j (cf. Figure 5.2) while guaranteeing the appropriate robot configuration to reach the next target T_{j+1} (cf. section 5.3.1), and so on.

Before we give more details of the navigation strategy based on sequential target reaching, let us introduce the used multi-controller architecture.

5.2.1 Proposed multi-controller architecture

To perform the navigation based on sequential target reaching, the multi-controller architecture depicted in Figure 5.3 is used. This architecture is composed of several blocks:

- The "Target assignment" block lets us obtain, at each sample time, the current waypoint (target) to reach. This block is detailed in subsection 5.2.2.

- The "Control law" block ensures asymptotic stability to reach the current assigned waypoint $T_j(x_{T_j}, y_{T_j}, \theta_{T_j}, v_{T_j})$. The details of this block are given in section 5.3.

- The "Obstacle avoidance" block is activated when an obstacle obstructs the

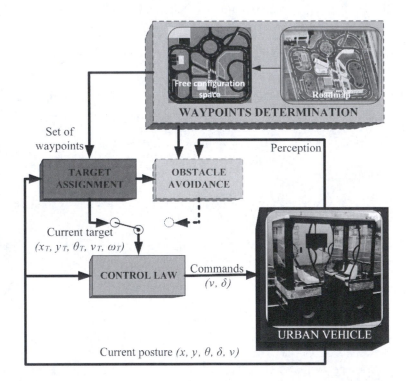

FIGURE 5.3 (See color insert) Proposed multi-controller architecture to perform autonomous vehicle navigation based on sequential target reaching.

robot's movement toward its current assigned waypoint. The used obstacle avoidance is based on the PELC technique (cf. section 2.3, page 31) and permits us to avoid locally any obstructing obstacle. This situation (obstacle avoidance) will be shown in section 5.5.2, where a group of VipaLab, performing a platooning (using a navigation through waypoints) have to avoid an obstructing obstacle.

- The "Waypoint determination" block (dashed green box in Figure 5.3) obtains the set of appropriate waypoints configuration. Section 5.4 focuses on the different developed techniques to obtain these waypoints.

5.2.2 Sequential target assignment

The strategy to assign, at each sample time, the waypoint to reach by the vehicle is shown in Algorithm 7. The stable and reliable control law defined in section 3.2.2, page 57, is used to reach each assigned waypoint while ensuring that the vehicle's trajectory is always within the road boundaries (cf. section 5.3.1).

The error conditions, E_{dis} and E_{angle}, are used to switch to the next waypoint when the vehicle's position is inside a circle given by the center (x_{T_j}, y_{T_j}) and a radius E_{dis}. Hence, the current waypoint index is updated with the next waypoint and the vehicle has to thereafter adapt its movement according to this new target. If the vehicle does not satisfy the distance and orientation error conditions (the errors d and $e_\theta >$ than E_{dis} and E_{angle} respectively) when crossing the Y_{T_j} axis (cf. Figure 5.2), then the vehicle must nevertheless switch to the next waypoint. Obviously, this situation should not occur if the environment is accurately modeled/identified and the control law well settled. Despite all these aspects, if this situation happens, then the value of the maximal distance and angular errors can be used to decide if the vehicle could or not continue its navigation. This fault detection/diagnosis is not addressed in this manuscript, but an accurate analysis of the used control law will be given in section 5.3.1 to determine the relation between the errors' upper bound and the used controller's parameters (cf. section 3.2.2, page 57).

It is also interesting to mention that the definition of Y_{T_j} axis, as in sections 2.3.2 (page 35) or 4.3.2 (page 83), guide the task achievement. In section 2.3.2 it is used to perform elementary obstacle avoidance and in section 4.3.2 to perform a trajectory planning algorithm based on PELC. This axis is used here as a mean to switch to the next waypoint.

5.3 CONTROL ASPECTS

As mentioned above, to obtain safe and reliable vehicle navigation, it is important to have an appropriate control law to sequentially reach the assigned waypoints. The control law defined in section 3.2.2, page 57 has been used in the presented control architecture (cf. Figure 5.3), to ensure the targeted navigation. It is to be noticed that this control law, as mentioned in section 3.2.2, is asymptotically stable, to reach any static or dynamic target. It allows in the current study to reach any assigned target

---

Algorithm 7: Sequential target assignment

Require: Vehicle pose, current target T_j and a set of N sorted waypoints

Ensure: Reaching T_j while guaranteeing to the vehicle to be in the best configuration to reach after the next waypoint T_{j+1}.

1: **if** (($d \le E_{dis}$ **and** $e_\theta \le E_{angle}$) **or** ($x^{T_j} \ge 0$))

    {x^{T_j} is the coordinate of the vehicle in the local Target frame $X_{T_j} Y_{T_j}$ (cf. Figure 5.2)}

    then

2:     Switch from the current target T_j to the next sequential waypoint T_{j+1}

3: **else**

4:     Keep going to waypoint T_j

5: **end if**

---

with a velocity (cf. equation 3.16) which can be $\ne 0$. This is important to insure smooth vehicle navigation between the different waypoints (cf. section 5.2).

Even if the used control law is proved asymptotically stable (using Lyapunov function analysis) [Vilca et al., 2015a], it does not allow us to obtain directly the error values (position and orientation) when the vehicle is in the immediate vicinity of the target to reach. This information is important guarantee the safety of the navigation. In fact, to insure safe navigation, the vehicle has to pass through each waypoint with maximum accuracy. It is supposed obviously that the waypoints were appropriately placed in the environment (cf. section 5.4 for accurate development of the waypoints planning phase).

The next subsection (5.3.1) aims to address the control law error estimation. Subsection 5.3.2 will address the way to enhance the smoothness of the switch between the waypoints.

5.3.1 Reliable elementary target reaching

The aim of this subsection is to determine the relation between the upper bound of the errors d and e_θ, denoted E_{dis} and E_{angle} (cf. Figure 5.2) and the controller parameters $\mathbf{K}$ (cf. section 3.2.2, page 57). The proposed analysis consists of determining the minimum distance d_{iMin} (cf. Figure 5.4) which allows us to satisfy at the same time, the vehicle's kinematic constraints (cf. section 3.8, page 58) and the final errors ($d_f \le E_{dis}$ and $e_{\theta_f} \le E_{angle}$) when the vehicle reaches the assigned target (at the final time $t = t_f$).

Figure 5.5 shows the global scheme to obtain the minimal initial distance d_{iMin}:

- e_{θ_0} and $e_{RT0} < \pi/2$, correspond to the initial stability conditions of the defined control law (cf. section 3.2.2, page 57).

- $\mathbf{K} = (K_d, K_l, K_o, K_x, K_\theta, K_{RT})$, is the constant vector gain, characterizing the control law parameters. $\mathbf{K}$ is fixed while taking into account the vehicle constraints such as maximal velocity v_{max} and minimum curvature radius $r_{c_{min}}$.

- E_{dis} and E_{angle} correspond to the upper error bounds to satisfy at $t = t_f$.

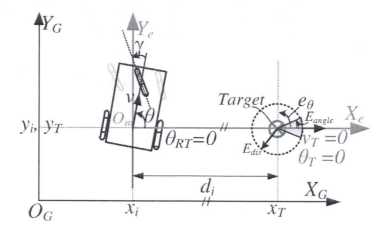

FIGURE 5.4 Limit vehicle's configuration for tuning d_{iMin} while taking into account the error bounds E_{dis} and E_{angle}.

To simplify the controller analysis, the orientation error e_θ and distance d are handled separately.

Firstly, the orientation error is computed considering enough initial distance of the vehicle to target d_i ($d_i \gg E_{dis}$) to permit monotonous convergence of e_θ toward zero (cf. section 3.2.2, page 57). This consideration permits us to estimate the minimum time to attain effectively $e_\theta \leq E_{angle}$. The following analysis considers a static target ($\dot{x}_T = \dot{y}_T = 0$ and $r_{cT} \to \infty$) and an extreme vehicle configuration, $| e_{\theta_0} | = \pi/2 - \zeta$ (where ζ a small positive value ≈ 0), hence the vehicle has initially the maximum admissible orientation error with respect to the target. The idea is to use the analysis of this limit in the vehicle's orientation error e_θ, which correspond to the slowest possible error convergence, to extrapolate thereafter the result for less

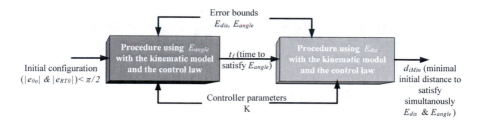

FIGURE 5.5 Block diagram of the analysis. In the left and right blocks, the procedures use, respectively, equations 5.3 and 5.5.

critical vehicle configuration $\mid e_{\theta_0} \mid \ll \pi/2$. Indeed, for less critical configuration, the convergence of e_θ will be faster than the limit defined case ($e_\theta \longrightarrow \pi/2$).

According to the developments given in [Vilca et al., 2015a] while considering the described limit configurations of the vehicle and the target, the analytic function of e_θ is obtained. This function approximates faithfully enough the evolution of the error e_θ when the vehicle is close to the target. It is given by:

$$e_\theta = f_\theta(t, \mathbf{K}, e_{\theta o}) = 2 \tan \left(\frac{e_{\theta o}}{2} \right) \left[\frac{(C + \cos(e_{\theta o}))(C - 1)}{(C - \cos(e_{\theta o}))(C + 1)} \right]^{C/2} e^{-\frac{K_x AB}{K_o} C^2 t_f}$$

$$(5.2)$$

where $A = K_d d$, $B = K_o K_\theta$ and $C = \sqrt{(A/B + 1)}$.

Using equation 5.2 and a fixed value of $\mathbf{K}$ it is immediate to compute the time t_f (cf. equation 5.3) necessary to obtain $e_\theta = E_{angle}$.

$$t_f = f_\theta^{-1}(E_{angle}, \mathbf{K}, e_{\theta o}) \mid_{e_{\theta o} = \pi/2 - \varsigma}$$

$$(5.3)$$

If $t > t_f$, then e_θ will be certainly $\leq E_{angle}$.

Secondly, once the time t_f is fixed, let us use it to determine d_{iMin} which permits us to guarantee always $d_f \leq E_{dis}$ and $e_{\theta_f} \leq E_{angle}$ for any initial vehicle configuration respecting: e_{θ_0}, $e_{RT0} < \pi/2$ and $d_i \geq d_{iMin}$. For a fixed navigation time t_f, the maximal possible initial distance d_i permitting us to reach the target is given when the vehicle's initial configuration corresponds to $e_{RT} = 0$ and $e_\theta = 0$ (straight line to the target). Obviously, the larger d_i is, the more certain we are that the vehicle in extreme configuration, as depicted in Figure 5.4, could reach the target with appropriate $e_\theta \leq E_{angle}$.

Taking $e_{RT} = 0$ and $e_\theta = 0$, the evolution of d can be written [Vilca et al., 2015a]:

$$d = d_i e^{-K_x K_d t}.$$

$$(5.4)$$

Therefore, using equation 5.4 and while knowing that the objective here is to have $d = E_{dis}$ at $t = t_f$, it could be easily concluded that:

$$d_{iMin} = E_{dis} e^{K_x K_d t_f}.$$

$$(5.5)$$

Some simulations, validating the above result, are given in subsection 5.3.3.1. More details about the different developments to obtain d_{iMin} are available in [Vilca et al., 2015a].

5.3.2 Smooth switching between targets

When the vehicle switches from one target to another (for instance from T_{j-1} to T_j as depicted in Figure 5.2), the value of controller variables $\mathbf{C}_v = (e_x, e_y, e_\theta, e_{RT}, v_T, r_{c_T})$ (cf. section 3.2.2, page 57) can change abruptly. These hard switches could induce, in certain situations, the actuators to jerk (v and γ (cf. equations 3.16 and 3.17)). This aspect could induce in certain applications, such transportation tasks, the discomfort of passengers.

It is proposed in this subsection to avoid this hard switch by introducing a smooth evolution of these control variables for a certain amount of movement distance d_s (smoothness distance) without obviously disturbing the vehicle's safe navigation [Vilca et al., 2013b]. This distance depends on initial distance (% of d_i) separating the vehicle from the next target T_{j+1}. A Sigmoid function is applied to the controller variables $\mathbf{C}_v$ along d_s. The new Smooth Virtual Controller variables ($\mathbf{SVC}_v$) are designed according to the covered distance $d_c = d_i - d$, where d is the current distance to the current target (cf. Figure 5.6). The $\mathbf{SVC}_v$ function is given by:

$$\mathbf{SVC}_v(d_c) = \mathbf{C}_{vi} + \frac{(\mathbf{C}_v - \mathbf{C}_{vi})}{1 + e^{-a(d_c - d_0)}} \tag{5.6}$$

where $\mathbf{C}_{vi}$ and $\mathbf{C}_v$ are, respectively, the initial and current values of the controller variables. For example, for e_x, an element of $\mathbf{C}_v$, when the target switches from T_j to T_{j+1}, e_{xi} is the value before switching to T_{j+1} and e_x is the current error w.r.t. the target T_{j+1}; d_0 is the value where the function has half of its current value and a is a constant value related to the slope of the sigmoid function. It is designed to attain the effective value ($\mathbf{SVC}_v \approx \mathbf{C}_v$) when $d_c = d_s$ (cf. Figure 5.6). The validation of this part, using simulation results, is given in subsection 5.3.3.2.

5.3.3 Simulation results (control aspects)

These simulations show the performances (stability, reliability and smoothness) of the proposed navigation strategy based on sequential target reaching.

5.3.3.1 Elementary target reaching (stability / reliability)

The first simulations focus on the features of the proposed control law (cf. section 3.2.2, page 57) to reach a desired final configuration (pose and velocity). For each

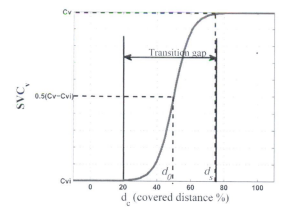

FIGURE 5.6 Evolution of the $\mathbf{SVC}_v$ used to ensure smooth control when target switching occurs.

simulation, the vehicle starts at the same position but with different initial orientations. Figures 5.7 (a) and (b) validate the analysis presented in subsection 5.3.1, where the minimum d_{iMin}, obtained for a limit vehicle configuration $e_\theta \approx \pi/2$, allows us to satisfy the bound of the errors for other, less critical, initial configurations. The desired final configuration is $(x_T, y_T, \theta_T) \equiv (15, 4, 0°)$ and $v_T = 1\ m/s$ (cf. Figure 5.7(a)).

The controller parameters $\mathbf{K} = (1/d_i, 0.6, 10, 0.1, 0.3, 0.01)$ were fixed in order to have fast and smooth vehicle trajectories and while taking into account the vehicle's maximal velocity $v_{max} = 1.5\ m/s$ and minimum radius of curvature $r_{c_{min}} = 3.8\ m$. While considering $E_{dist} \leq 0.1\ m$ and $E_{angle} \leq 5°$ and using equations 5.3 and 5.4 lead to $t_f \approx 10.5\ s$ (cf. equation 5.3), the minimum authorized initial Euclidean distance to the target is finally obtained $d_{iMin} = 10.6\ m$.

Figures 5.7 (a) and (b) show respectively the trajectory of the vehicle for different initial orientations and orientation errors. Figure 5.7(a) shows that the convergence of the system depends on the initial orientation error. Figure 5.7(b) shows that the errors are bounded (cf. equation 5.2) (black bold lines) and converge always to zero (cf. subsection 5.3.1). The Lyapunov function evolutions, for each of the above simulations, are shown in Figure 5.8(a) and confirm the asymptotic stability of the used control law. Furthermore, Figure 5.8(b) shows, as indicated, the evolution of the three terms composing the Lyapunov function (cf. equation 3.15, page 59) where the first term is $0.5 K_d d^2$, the second term is $0.5 K_l d_l^2$ and the third term is $K_o[1 - \cos(e_\theta)]$.

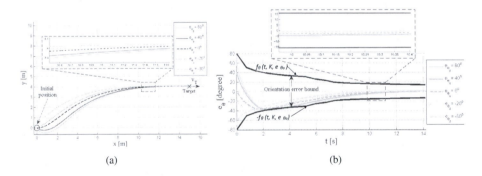

(a) (b)

FIGURE 5.7   (a) Trajectories of the vehicle for several initial orientations. (b) orientation errors (e_θ) for several initial orientations.

5.3.3.2 Sequential target reaching/switching

The following simulations focus on the reliability and the smoothness of the proposed navigation strategy based on sequential target reaching. The proposed control law (cf. section 3.2.2, page 57) will also be evaluated to reach/track static as well as dynamic targets. Figure 5.9(a) shows the trajectories of the vehicle for sequentially reaching several static targets (T_i, $i = 1, \ldots, 6$) and tracking after a dynamic target T_d (sinusoidal trajectory). The static targets are positioned at different initial distances d_i

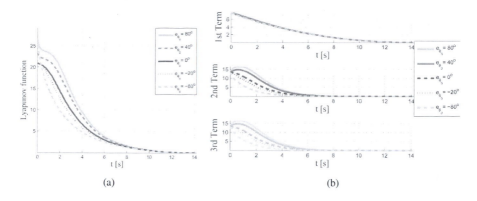

FIGURE 5.8 (a) Lyapunov function values for several initial orientations. (b) different terms of the Lyapunov function (cf. equation 3.15, page 59).

and orientation angles between them ($45°$ until T_5 and $0°$ for T_6). The velocity profile of the targets for each simulation are $v_T = 0.1,\ 0.5,$ and $1.0\ m/s$, respectively. The values of the controller parameters are $\mathbf{K} = (1/d_i, 1.8, 8, 0.15, 0.6, 0.01)$ (d_i is the initial distance to the target). These parameters were chosen to obtain safe and smooth trajectories, fast response and velocity values within the limits of the vehicle, which are $v_{max} = 1.5\ m/s$ and $r_{c_{min}} = 3.8\ m$ ($\gamma_{max} = \pm19°$). It is to be noted that the vehicle converges to each assigned target (static and dynamic), located in different positions and with a different set of velocities. The dynamic target starts its movement when the vehicle reaches the last static target T_6.

Figure 5.9(b) shows the values of errors d and e_θ for the different targets to reach. For static targets ($T_i, i = 1, \ldots, 6$), the obtained values of errors just before switching from target T_i to T_{i+1} are shown. For dynamic targets, the evolution of $d(t)$ and $e_\theta(t)$ during the tracking phase are shown. It is observed that the distance and orientation errors of static targets depend on initial configuration (distance and orientation) and increase when the static targets are closer. As expected, for dynamic targets, the small target velocity profile has faster convergence toward zero.

The use of the sigmoid function is observed in the vehicle commands (velocity and steering angle) (cf. Figure 5.10) for the static targets with profile velocity of $v_T = 0.5$. It is noted that the sigmoid function contributes to avoid peaks at the transition phase and permits smoother vehicle commands while maintaining the stability of the control.

To highlight the genericity of the proposed navigation strategy based on successive waypoints, the application of this strategy was investigated when the waypoints are very close, as if the vehicle has to follow a trajectory. For that purpose, two sets of waypoints, selected from a reference trajectory, are used. The first set has a distance between waypoints equal to $2\ m$ and the other equal to $4\ m$ (cf. Figure 5.11). Figures 5.11 and 5.12 show respectively the vehicle's trajectories and the lateral E_{Lateral} and angular e_θ errors w.r.t. the reference trajectory (for the two set of waypoints). It

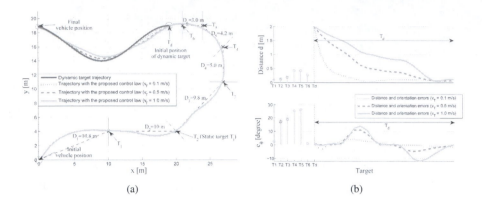

(a)   (b)

FIGURE 5.9 (a) Trajectories of the vehicle for several target velocities. (b) distance and orientation errors of the vehicle for several targets' pose/velocities.

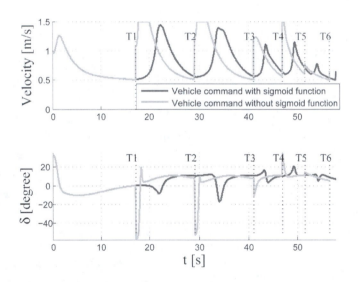

FIGURE 5.10 Control commands with and without adaptive sigmoid ($\mathbf{SVC}_v$) use.

can be noted that the obtained vehicle trajectories are close enough to the reference trajectory; and as expected, the lateral and angular errors are smaller when the fixed distance between the waypoints decreases. Therefore, the proposed navigation strategy and control law permit accurate trajectory tracking behavior if the waypoints are close enough.

It is to be mentioned also that in [Vilca et al., 2015a], an interesting comparison between the proposed control law and those proposed in the literature which are dedicated to path following or trajectory tracking (e.g., [Samson, 1995] [Daviet and Parent, 1997] or [Siciliano and Khatib, 2008]). The obtained results confirm that even if the proposed control law has not been designed explicitly to take into account a reference path/trajectory, the obtained results are very satisfactory [Vilca et al., 2015a]. Moreover, the advantage of the proposed control law is its flexibility to perform autonomous navigation. Indeed, the proposed control law needs only to know the current pose and velocity of the target instead of the entire trajectory to track.

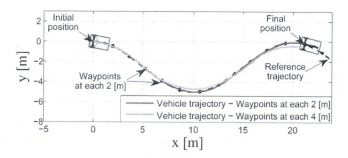

FIGURE 5.11   Vehicle trajectories for different distances between waypoints.

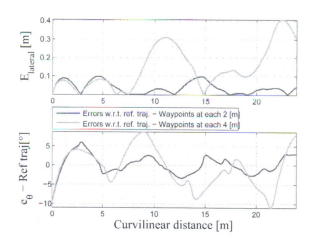

FIGURE 5.12   Errors w.r.t. reference trajectory for different distances between waypoints.

5.4 WAYPOINTS' CONFIGURATION ASPECTS

Once the principle of navigation strategy, based on sequential target reaching, is validated in terms of control stability and smoothness (cf. section 5.3), let us address in this section the means to obtain the most appropriate configuration (number, postures, etc.) of these waypoints in the environment. The aim is to ensure, in all cases, navigation safety, but in addition, the navigation's smoothness and rapidity can be taken into account.

It is presented in what follows two methods to obtain this appropriate set of waypoints. The first (cf. section 5.4.2), which is the most complete, uses a multi-criteria function to obtain the optimized waypoints set. The second, which is more restrictive, use initially an existing trajectory and selects the minimum number of waypoints belonging it (cf. section 5.4.3). Before we give the details of these methods, it is given in what follows, the state of the art concerning waypoints planning.

5.4.1 State of the art

Different algorithms can be used to obtain waypoints configuration such as A^*, D^* [Choset et al., 2005], Rapidly Random Tree (RRT) [Kuwata et al., 2008], Sparse A^* Search (SAS) [Szczerba et al., 2000]. Configuration space (C-space), space of all possible configurations of the vehicle [Siciliano and Khatib, 2008], enables the identification of the safe area where the vehicle can navigate without a collision risk (free space C-space$_{free}$). C-space is used to compute the minimum distance to C-space$_{obst}$ (obstacle or road boundaries space). Figure 5.13 shows the C-space and its Voronoï diagram [Latombe, 1991] in grayscale w.r.t. the distance to the closest C-space$_{obst}$ (the whitest area represents the safest area). Typically, algorithms based on a grid map (e.g., A^* or D^*) produce the shortest path by optimization of a criterion such as the distance to the goal, distance to the risk area, etc. [Choset et al., 2005]. The algorithm begins generally at the final cell (final position) and traverses the cell's neighbors until it reaches the initial position. The cost of traveling through the neighbor is added to the total cost, the neighbor with the lowest total cost is selected, and so on. The process terminates once the initial position is reached. The path is given through the cell positions of the grid map while backtracking the cells which have the lower path cost, and sometimes a polynomial interpolation is used to obtain a smooth path [Connors and Elkaim, 2007]. In [Ziegler et al., 2008], the authors present an A^* algorithm using clothoid trajectories assuming constant velocity along them. Therefore, appropriate waypoints can be selected from this shortest path while only considering the cells where an orientation change occurs (w.r.t. its predecessor). Nonetheless, this algorithm does not consider former/initial vehicle orientation or its kinematic constraints.

Instead of using a grid map, it is possible also to obtain safe, feasible and smooth path using expanding tree algorithms (e.g., RRT, RRT* or SAS [LaValle, 2006], [Kuwata et al., 2008], [Karaman and Frazzoli, 2011] and [Szczerba et al., 2000]). This could be done by providing to the vehicle model the commands to reach the successive selected nodes until the goal [LaValle, 2006], [Kuwata et al., 2008] and

[Szczerba et al., 2000]. The basic process of RRT consists of selecting, at each sample time, a random node q_{random} in the C-space$_{free}$. This selection considers generally only position $q_{random} = (x_{random}, y_{random})$ without any *a priori* final vehicle orientation [LaValle, 2006]. Then, the commands (discrete values) are applied to the vehicle (from its current position and orientation) during a constant time t_{exp}. The vehicle model and constant commands allow us to predict the final vehicle position at the end of t_{exp}. The commands that produce the closest position q_{chosen} (a node which optimizes a dedicated task criterion [Vaz et al., 2010]) to current random node q_{random} are selected and stored with q_{choose}. A new expansion starts to reach q_{random} or to select a new random node q_{random}^{new}. Therefore, the waypoints can be selected, as in the case of a grid map, while only considering the nodes where an orientation change occurs (w.r.t. its predecessor node). Algorithms based on RRT are very useful for motion planning because they can provide the commands (based on the kinematic/dynamic model of the vehicle) to reach the desired final configuration [Kuwata et al., 2008] and [Vaz et al., 2010]. Moreover, it avoids the use of grid maps that can increase the computation time for large environments. In [Szczerba et al., 2000], the authors use the expanding tree for trajectory planning introducing different constraints such as maximum turning angle and route distance. Nevertheless, this method does not consider the vehicle's movements along the trajectory or localization uncertainties. In [Kallem et al., 2011], sequential composition of controllers (e.g., go to the landmark and wall following controller) are used to generate valid vehicle states q_{choose} to the navigation function. This approach avoids finding a single globally attractive control law and allows us to use some additional sensing capability of the vehicle when the landmark is unreachable (e.g., GPS-denied area). However, the obtained navigation function has complex computational processing. The most important drawbacks of expanding tree algorithms are the slow convergence to cover all space to reach the goal and in most cases it does not provide the shortest path since the nodes are randomly selected [Abbadi et al., 2011]. Furthermore, it is important to underline that in the RRT the control commands are maintained during a certain time, whereas in the presented work (cf. section 5.4.2), the vehicle's movement takes into account the definition of the used control law in addition to the vehicle model. A comparison with RRT and Voronoï approaches is shown in subsection 5.4.4.1.

It is proposed mainly in what follows, a method based on expanding tree to obtain

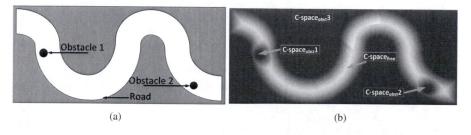

(a)   (b)

FIGURE 5.13   (a) Road scheme and (b) its C-space and Voronoï diagram.

the optimal waypoint configuration in a structured environment (as shown in Figure 5.1). It allows us to consider constraints such as the kinematic model and the used control law. Criteria to optimize are related to the vehicle's kinematic constraints (non-holonomy, maximum velocity and steering angle) and localization uncertainties.

5.4.2 Optimal Multi-criteria Waypoint Selection based on Expanding Tree (OMWS-ET)

This section aims to present a method of waypoints selection in a structured environment in order to perform safe and flexible vehicle navigation. The waypoints are obtained using an Optimal Multi-criteria Waypoint Selection based on Expanding Tree (OMWS-ET). Another OMWS-based Grid Map (OMWS-GM) has been proposed in [Vilca et al., 2015b] but will not be detailed in what follows. In both proposed optimizations, waypoints are selected considering safe position on the road; feasibility of trajectories (smooth changes between the successive points and respecting the vehicle's kinematics constraints) and system uncertainties (modeling / localization). The waypoints selection approaches are formulated as an optimization problem and solved using dynamic programming [Bellman, 1957, Bellman et al., 1959]. A generic formulation is given below.

Optimization problem formulation: *For each discrete state $x_k \in X$ where X is a nonempty and finite state space, the objective is to obtain the sequence of states to reach the final state x_K while minimizing the following cost function:*

$$C(x_K) = \sum_{k=1}^{K} g(Pred_{x_k} \rightarrow x_k) + h(x_K) \qquad (5.7)$$

where $Pred_{x_k}$ is the predecessor state of x_k; g is the immediate traveling cost function to go from $Pred_{x_k}$ to x_k; h is the future traveling cost function (heuristic) to go from the current state to the final state x_K. When the current state is the final state x_K then $h(x_K)$ is equal to zero. This function h contributes to improve the convergence of the sub-optimal solutions toward the global optimal one [Bertsekas, 1995].

Before giving more details about OMWS-ET, let us present the definition of the used expanding tree. The expanding tree is composed of nodes and edges which have the following properties:

- Each node q_j is defined by the pose of a waypoint given by $(x_{q_j}, y_{q_j}, \theta_{q_j})^T$, one predecessor node q_i (except for the initial node) and traveling cost values $g(q_j)$ and $h(q_j)$ (cf. equation 5.7).

- Each edge ξ_{ij} corresponds to the link between q_i to q_j nodes.

- $g(q_i \rightarrow q_j) = g(\xi_{ij})$ is the traveling cost from q_i to q_j.

- $h(q_j) \in [0, 1]$ is the heuristic traveling cost from the current node q_j to the final node (final vehicle pose). It is defined according to the Euclidean distance d_{q_j} from the position (x_{q_j}, y_{q_j}) to the final targeted vehicle pose by:

$$h(q_j) = k_h \left(1 - e^{-d_{q_j}/k_e} \right) \tag{5.8}$$

where $k_h \in [0, 1]$ enables us to tune the significance of $h(q_j)$ in the total cost function (5.7). The exponential function was chosen because it gives values between 0 and 1 for positive values of d_{q_j}. The constant $k_e \in \mathbb{R}^+$ is used to scale the value of d_{q_j} according to the dimensions of the environment. The value of $h(q_j)$ (5.8) decreases as the next possible selected node gets closer to the final pose.

The traveling cost $g(\xi_{ij}) \in [0, 1]$ is designed to obtain an appropriate balanced safe, smooth, feasible and fast vehicle trajectory. It is defined as:

$$g(\xi_{ij}) = k_1 \bar{w}_j + k_2 \Delta \bar{v}_{ij} + k_3 \Delta \bar{\gamma}_{ij} + k_4 \Delta \bar{e}_{l_{ij}} \tag{5.9}$$

where k_1, k_2, k_3 and $k_4 \in \mathbb{R}^+$ are constants which are defined by the designer to give the right balance (according to the navigation context, e.g., focus more on safety with regard to smoothness) of each term of the criteria (cf. equation 5.9). To normalize the traveling cost, $k_i | i = 1, .., 4$ must satisfy:

$$k_1 + k_2 + k_3 + k_4 = 1. \tag{5.10}$$

The normalization of the individual criterion given in equation 5.9 allows us to simplify the choice of k_i to select the priority of a term w.r.t. the others according to the navigation context.

The first term of the cost function (cf. equation 5.9) is related to the safety of the navigation (5.11). The second and third terms are respectively related to the speed (cf. equation 5.12) and smoothness (cf. equation 5.17) of the trajectory. The fourth term is related to the feasibility of the vehicle trajectory while considering localization uncertainties, i.e., the risk of colliding with an obstacle while considering inaccuracies in the vehicle position and orientation. This last term allows considering the kinematic model of the vehicle and the control law. The detail of each term is given in the following:

- The term $\bar{w}_j \in [0, 1]$ is related to the distance from the node q_j to the closest C-space$_{obst}$. It is given by equation 5.11, the normalized distance $d_{q_j\_To\_Obst}$ to the closest C-space$_{obst}$, and it is given by:

$$\bar{w}_{q_j} = 1 - \frac{d_{q_j\_To\_Obst}}{d_{max\_To\_Obst}} \tag{5.11}$$

where $d_{max\_To\_Obst}$ is the maximum value among all $d_{q_i\_To\_Obst}$ of all cells in the C-space$_{free}$. As an example, Figure 5.14 shows a discretized environment using a grid map; 3 cells localized at (a, b), (i, j) and (m, n) are emphasized. $d_{max\_To\_Obst}$ is equal, in this example, to the maximum distance $d_{mn\_To\_Obst}$.

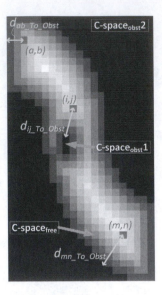

FIGURE 5.14 Representation in grayscale w.r.t. the distance to the closest C-space$_{obst}$ (the whitest area represents the safest one).

- The term $\Delta\bar{v}_{ij} \in [0,1]$ is related to the velocity from q_i to q_j, v_{ij}. It is given by:

$$\Delta\bar{v}_{ij} = 1 - \frac{v_{ij}}{v_{max}} \qquad (5.12)$$

where v_{max} is the maximum velocity of the vehicle. We estimate v_{ij} as a function of the curvature of the expected trajectory. The maximum velocity occurs when the curvature is zero (straight line) and the minimum velocity $v_{min} \neq 0$ occurs when the curvature is bigger than the value corresponding to γ_{max} (the maximum vehicle steering angle, (cf. section 3.2.2, page 57)). This consideration allows the vehicle to maneuver without a risk of collisions while enhancing the passenger comfort (for instance, to limit the centripetal forces [Labakhua et al., 2008]). The minimum and maximum values of velocity and steering angle are defined by the designer according to the vehicle characteristics. The curvature is estimated using the orientation of the current node and its predecessor. Hence, v_{ij} is computed as:

$$v_{ij} = v_{max} - \Delta\bar{\theta}_{ij}(v_{max} - v_{min}) \qquad (5.13)$$

where $\Delta\bar{\theta}_{ij} \in [0,1]$ is the normalized estimated curvature related to the variation of orientation between the current node q_j and its predecessor q_i. It is defined as

$$\Delta\bar{\theta}_{ij} = \frac{|\theta_j - \theta_i|}{\Delta\theta_{max}} \qquad (5.14)$$

where $\Delta\theta_{max}$ is the maximum possible variation between a probable orientation of the current node w.r.t the orientation of its predecessor. This value is defined according to the steering capability of the vehicle. θ_j and $\theta_i \in]-\pi, \pi]$ are computed using the position of the current node $q_i(x_i, y_i)$, its predecessor $q_m(x_m, y_m)$ and its probable successor $q_j(x_j, y_j)$. They are given by

$$\theta_i = \arctan\left([y_m - y_i]/[x_m - x_i]\right) \tag{5.15}$$
$$\theta_j = \arctan\left([y_i - y_j]/[x_i - x_j]\right). \tag{5.16}$$

• The term $\Delta\bar{\gamma}_{ij} \in [0, 1]$ is related to the variation of steering angle along the vehicle trajectory from q_i to q_j (for instance, Figure 5.15 shows a vehicle trajectory between two nodes). It is given by

$$\Delta\bar{\gamma}_{ij} = \frac{\sum_{q_i}^{q_j} |\Delta\gamma_{ij}|}{n_{q_{ij}}\gamma_{max}} \tag{5.17}$$

where $n_{q_{ij}}$ is the considered point number of the vehicle trajectory between q_i and q_j, and γ_{max} is the maximum steering angle of the vehicle. This term $\Delta\bar{\gamma}_{ij}$ computes the sum of the $\Delta\gamma_{ij}$ to obtain the total variation of the steering angle along the vehicle trajectory. $\Delta\bar{\gamma}_{ij}$ uses the kinematic model and the control law to estimate the vehicle's trajectory.

• The term $\Delta\bar{e}_{l_{ij}} \in [0, 1]$ is the normalized maximum deviation of the vehicle's trajectory w.r.t. the straight line that joins the positions (x_q, y_q) of q_i and q_j (cf. Figure 5.15). It is computed as

$$\Delta\bar{e}_{l,j} = \frac{|\Delta e_{l_{ij}}|}{max\{\Delta e_l\}} \tag{5.18}$$

where $max\{\Delta e_l\}$ is the maximum deviation of all trajectories from the node q_i to the node q_j while considering the position and orientation uncertainties (ϵ_d and c_θ respectively given in Figure 5.15). This term $\Delta\bar{e}_{l,j}$ allows us to estimate the collision risk using the vehicle trajectory that takes into account the kinematic model, the control law and localization uncertainties (position and orientation). Figure 5.15 shows an illustration where the vehicle has an ellipse of localization uncertainties with axes ϵ_d^l and ϵ_d^t. The vehicle trajectories start at $\pm\epsilon_d^l$ in lateral distance (longitudinal distance is set to 0), and $\pm\epsilon_d^t$ in longitudinal distance (lateral distance is set to 0) from the vehicle position with a $\pm\epsilon_\theta$ from the vehicle presumed orientation, i.e., we consider all extreme configurations to obtain, according to these maximum error configurations, the maximum lateral deviation (Δe_l). The trajectories are obtained using the kinematic model and the used control law in an offline simulated procedure.

Algorithm 8 shows in pseudo code, the proposed method which uses expanding tree to obtain the optimal waypoints configurations w.r.t. the optimized multi-criteria function (5.9). Figure 5.16 shows the first steps of the algorithm where, for instance,

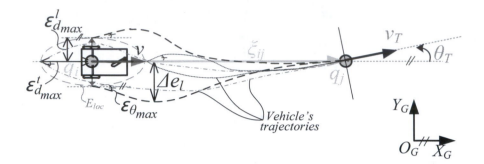

FIGURE 5.15 Vehicle's trajectories which start from extreme configurations $(\pm\epsilon^l_{d_{max}}, \pm\epsilon^t_{d_{max}}$ and $\pm\epsilon_{\theta_{max}})$ in the localization uncertainties ellipse E_{loc}. Δe_l is the maximum lateral deviation of all vehicle trajectories.

the branch number of each node is $n_t = 3$ and each branch orientation w.r.t. the vehicle orientation is given by:

$$\alpha = \pm i \Delta\alpha, \quad i = \begin{cases} 0, 1, \ldots, (n_t - 1)/2; & \text{if } n_t \text{ is odd} \\ 1, 2, \ldots, n_t/2; & \text{if } n_t \text{ is even} \end{cases} \quad (5.19)$$

where $\Delta\alpha$ is a constant angle defined according to the vehicle characteristics.

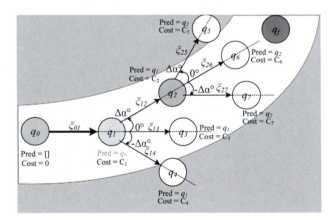

FIGURE 5.16 Expanding tree method to obtain the appropriate set of way-points.

The edge distance ξ is the Euclidean distance between two successive nodes and it depends on the environment dimensions, e.g., if the environment has a narrow passage, then ξ must cope with this dimension. We consider that the edge orientation is the vehicle orientation at the current node position (cf. Figure 5.16). Thus, at the

---

Algorithm 8: Waypoint selection based on expanding tree

Require: Initial pose p_i, final pose p_f, branch number n_t, edge distance ξ, branch
      orientation $\Delta\alpha$, tolerable error distance ϵ and C-space$_{free}$
Ensure: Set of waypoints S_p
  1: Init the initial node $q_0 = p_i$, $g_0 = 0$ and $Pred_{q_0} = \oslash$
  2: Init the current node to expand $q_i = q_0$
  3: Init $Tree(q_i)$ =Expansion_Tree (cf. Algorithm 9) with $\alpha = 0$ {Initial expansion}
  4: Set the new node to expand $q_i = r_t$ where $r_t \in Tree(q_i)$
  5: Set $Pred_{r_t} = q_i$ and compute the total cost $C(r_t)$ (cf. equation 5.7)
  6: **while** $|q_i - p_f| < \epsilon$ **do**
  7:    Compute the $Tree(q_i) =$ Expansion_Tree
  8:    {refers to Algorithm 9 with the set of $\alpha = \pm i\Delta\alpha$}
  9:    **for** $r_t \in Tree(q_i)$ **do**
10:      **if** $r_t \in$ C-space$_{free}$ **then**
11:        Compute the total cost $C(r_t)$ (cf. equation 5.7)
12:        $Pred_{r_t} = q_i$
13:        Add r_t to the queue Q
14:      **end if**
15:    **end for**
16:    Sort the queue Q in ascending order of total cost C
17:    Get the first value of queue $q_i = Q(first)$ and remove it from Q
18: **end while**
19: S_p is the set of predecessor nodes of $q_i = p_f$.

Algorithm 9: Expansion_Tree

Require: Current node q_i, set of α $S(\alpha)$, edge distance ξ
Ensure: Nodes of $Tree(q_i)$
  1: Init $Tree(q_i) = \oslash$
  2: **for** $\alpha_t \in S(\alpha)$ **do**
  3:    Compute the orientation $\theta_{r_t} = \theta_{q_i} + \alpha_t$
  4:    Compute pose $r_t = q_i + [\xi\cos(\theta_{r_t}), \xi\sin(\theta_{r_t}), \alpha_t]^T$
  5:    Add r_t to $Tree(q_i)$
  6: **end for**

---

beginning the first expansion of q_0 is given with $\alpha = 0$ because the vehicle starts
at initial fixed pose (cf. line $3 - 5$ of Algorithm 8). This initial expansion is made
to respect the kinematic constraints where the rotation of the vehicle requires a displacement (linear velocity $\neq 0$) of the vehicle. Therefore, the successive node q_1 has
a different possible orientation and so on (cf. Figure 5.16). The algorithm selects the
node which has the lower total cost $C(q_j)$ (cf. equation 5.7). When two or more nodes
have the same cost, the algorithm selects the last saved node. Figure 5.16 shows the
successive steps, and the node q_2 was selected from the expansion of q_1 $\{q_2, q_3, q_4\}$,
which has the lower total cost value. The set of waypoints is obtained while tracking
the predecessor nodes of the nodes with lower total cost. The selection of the node
with lower total cost (cf. Algorithm 8, line 16-17) allows us to avoid the deadlock

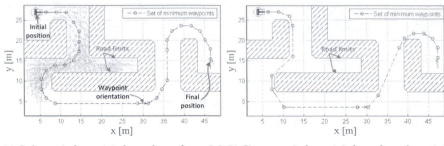

(a) Safest path: $k_1 = 1.0$, $k_2 = k_3 = k_4 = 0.0$ (b) Shortest path: $k_2 = 1.0$, $k_1 = k_3 = k_4 = 0.0$
and $k_h = 0.1$ and $k_h = 0.1$

FIGURE 5.17 Set of waypoints for different parameter values k_i of the traveling cost.

areas because the successive branches from the nodes in this deadlock area will be in
C-space$_{obst}$ (cf. Figure 5.17(a)).

The smoothness of the vehicle trajectory depends on the number of branches of
each tree n_t, maximum branch orientation $\alpha_{max} = n_t \Delta\alpha/2$ and edge distance ξ.
The drawback of using a large number of n_t is the increase of the processing time
required to obtain the set of waypoints. The vertex distance ξ is set to detect obstacles
between the successive nodes.

As described above, the traveling cost (cf. equation 5.9) depends on four param-
eters ($k_i | i = 1, \ldots, 4$, which satisfy equation 5.10) related respectively to the safety,
velocity, less steering and taking into account uncertainties. The values of these parameters are fixed according to the desired navigation and environment conditions.
A pragmatic procedure to set these parameters consists first in identifying the main
desired vehicle behavior and setting its parameter k_i with a value greater than 0.5
(cf. Figure 5.17). The other parameters will be tuned according to the designer's secondary priorities. Figure 5.17 shows the set of waypoints when only the term with
highest priority is considered in the traveling cost function. For instance, in Figures
5.17 (a) and (b), the priority is given respectively to the safest and the shortest paths.
More examples of different tuned parameters of k_i and the actual potentialities of the
proposed OMWS-ET will be shown in sections 5.4.4 and 5.5.1.

5.4.3 Waypoints selection based on existing safe trajectory

To simplify obtaining the set of waypoints, this one could be selected from an
already existing path. Indeed, a safe reference path can be obtained easily using
different methods such as a Voronoï diagram [Latombe, 1991] or potential fields
[Khatib, 1986]. Nevertheless, adding this step of path planning restricts considerably
the C-space$_{free}$ to only a curvilinear line (corresponding to the used path). Thus, the
optimality of the obtained set of waypoints (cf. section 5.4.2) is not guaranteed at
all. It is also interesting to mention that a path could also be obtained using an actual
recorded vehicle trajectory [Vilca et al., 2013b].

Different criteria can be considered to obtain the minimum number of straight lines that closely fit the reference path. Criteria such as the Euclidean or curvilinear distance, orientation or radius of curvature between waypoints can be used to fix the desired waypoints on the path. The proposed waypoint selection given in Algorithm 10 is maximally simplified. The discretized reference path $\mathbf{r}$ is composed of sorted positions $r_i = (x_{r_i}, y_{r_i})$ and its tangent orientations θ_{r_i}. The minimum number of straight line segments over the defined path is then computed while considering a constant threshold $\Delta_{\alpha_{max}}$ for the orientation variation of the path Δ_α (cf. Algorithm 10). Figure 5.18 shows one vehicle trajectory and the obtained waypoints using Algorithm 10 with $\Delta_{\alpha_{max}} = 5°$, $15°$, and $30°$, respectively. As expected, the switch between waypoints is smoother with a small value of $\Delta_{\alpha_{max}}$.

5.4.4 Simulation results (waypoints configuration aspects)

This section shows the proposed OMWS-ET features to lead to the optimal set of waypoints, obtained according to the environment characteristics and/or the task to achieve. Simulations given below use the VipaLab vehicle with maximum velocity $1.5\ m/s$, maximum steering angle $\gamma_{max} = \pm30°$, and maximum linear acceleration $1.0\ m/s^2$. The controller parameters are set to $\mathbf{K} = (1, 2.2, 8, 0.1, 0.01, 0.6)$ (cf. section 3.2.2, page 57). These parameters were chosen to obtain a good balance between accurate and fast response and smooth trajectory while taking into account the limits of the vehicle structural capacities. It is considered that the sample time is $0.01\ s$ and a maximum number of iterations is $n_I = 5000$ to stop OMWS-ET, when no solution can be obtained.

It will be shown in what follows that the comparison between OMWS-ET and the well-known RRT* algorithm (cf. subsection 5.4.4.1); the flexibility of OMWS-ET for local replanning when unexpected obstacles are detected (cf. subsection 5.4.4.2) and finally, the extension of OMWS-ET to multi-vehicle formation (cf. subsection 5.4.4.3). It is noted that in [Vilca et al., 2015b], more simulations were given, show-

---

Algorithm 10: Waypoint selection based on existing reference path

---

Require: Reference path $\mathbf{r} = (\mathbf{x}_r, \mathbf{y}_r)$ and $\Delta_{\alpha_{max}} \in \mathbb{R}^+$
Ensure: Set of waypoints S_p
  1: Init $j = 0$, $r_{w_j} = r_0$ (initial position of $\mathbf{r}$) and $\theta_{w_j} = \theta_{r_0}$ (tangent of the point along trajectory $\mathbf{r}$)
  2: **for** $r_i \in \mathbf{r}$ (sorted set of trajectory points) **do**
  3:     Compute $\Delta_\alpha = |\theta_{r_i} - \theta_{w_j}|$
  4:     **if** $\Delta_\alpha \geq \Delta_{\alpha_{max}}$ **then**
  5:         $j = j + 1$
  6:         Set $r_{w_j} = r_i$ and $\theta_{w_j} = \theta_{r_i}$
  7:         Add $w_j(r_{w_j}, \theta_{w_j})$ to S_p
  8:     **end if**
  9: **end for**

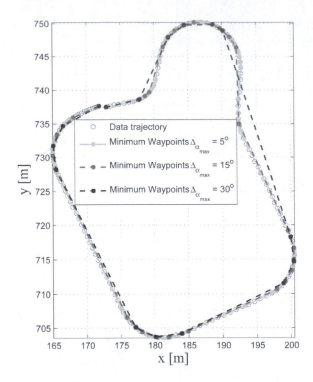

FIGURE 5.18 Example of waypoint selection based on a reference path and Algorithm 10.

ing for instance the comparison between grid map and expanding tree algorithms or between deterministic and probabilistic expanding tree.

As first simulations, Figures 5.19 (a) and (b) show, respectively, an example of the use of the proposed OMWS-ET (cf. Algorithm 8) and the reduction of the number of waypoints while using Algorithm 10 [Vilca et al., 2015a].

5.4.4.1 OMWS-ET versus RRT*

To highlight the advantages and the flexibility of the proposed OMWS-ET, a comparison with the popular RRT* algorithm [Karaman and Frazzoli, 2011] is presented in this subsection. The RRT* is based on the RRT (Rapidly-exploring Random Tree) already described in section 5.4.1 with an addition of the rewiring function which enables to reconnect the nodes to ensure that the edges have the path with minimum total cost. RRT* provides thus an optimal solution with minimal computational and memory requirements [Karaman and Frazzoli, 2011]. Moreover, RRT* is a sampling-based algorithm and the optimal solution depends on the number of iterations of the RRT* algorithm, i.e., the larger the number of iterations (more samples in the C-space$_{free}$), the closer is the effective optimal global solution. Therefore, to

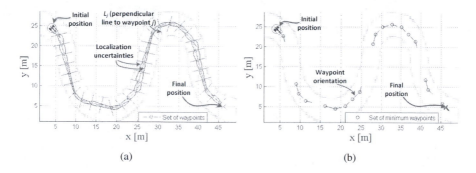

FIGURE 5.19 (a) Set of obtained waypoints using Algorithm 8 based on OMWS-ET and (b) minimum set of waypoints obtained by Algorithm 10.

compare the solutions obtained by the OMWS-ET with those obtained by the RRT* a few modifications in Algorithm 8 were made. The line 6 of Algorithm 8 was changed to a for loop from 0 to the maximum iteration number and the selection of the final pose at each iteration is obtained by sampling in C-space$_{free}$ (q_{random}), as the RRT* algorithm [Karaman and Frazzoli, 2011]. It is to be noted that q_{random} corresponds to a random sample (position) from a uniform distribution in the C-space$_{free}$.

To compare the two algorithms (OMWS-ET and RRT*), the safest obtained solution (which maximizes the distance to the environment border) is used as a criterion. Therefore, the parameters of the cost function of OMWS-ET (cf. equation 5.9) are fixed to $k_1 = 1.0$, $k_2 = k_3 = k_4 = 0.0$ and $k_h = 0.1$. In addition, the other parameters are fixed as the branch number $n_t = 5$; the edge distance $\xi = 2.5\ m$ and $\Delta\alpha = 15°$. The RRT* algorithm described in [Karaman and Frazzoli, 2011] was also modified to obtain a cost function according to the safety $\bar{w}_i$ (distance to the border) instead of the Euclidean distance between nodes. The vehicle kinematic model with constant linear velocity and steering angles ($v = 1.0\ m/s$ and $\gamma = -15, -7.5, 0, 7.5, 15°$) respectively, during $t_{exp} = 2.5\ s$ was used to produce the new nodes of the RRT*. Figure 5.20 shows the obtained path solutions according to RRT*, OMWS-ET and Voronoï [Latombe, 1991] algorithms. The Voronoï obtained path (cf. Figure 5.20(c)) is given only because it is the best reference w.r.t. the adopted comparison criterion (safety criterion). Indeed, the Voronoï path enables to always obtain the safest possible path [Latombe, 1991]. It can be noted that the two obtained paths using RRT* and OMWS-ET are generally close enough and far enough from the way border (cf. Figures 5.20(a) and 5.20(b)). Important differences are nevertheless observed in the obtained final results (cf. Figure 5.20(c)). In fact, the obtained set of waypoints using RRT* are closer to the border which is due to the fact that RRT* expands its branches while adopting constant commands (v, γ, t_{exp}). These constant commands generate the next nodes with only a single possible orientation (for each node). Contrary to that, in the proposed OMWS-ET, each new obtained node q_j has different possible orientations and velocities. Thus, for the

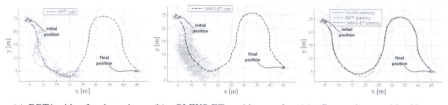

(a) RRT* with a few branches   (b) OMWS-ET with a few (c) Comparison with Voronoï
                                branches   path

FIGURE 5.20   Three obtained paths according to Voronoï, RRT* and OMWS-ET.

same position, many more possible vehicle states (different orientations and velocity set-points) are taken into account in the optimization process.

Table 5.1 shows different performance criteria to compare the obtained path. It is shown that the obtained path based on OMWS-ET is closer than the RRT* to the optimal obtained solution using Voronoï methodology. It validates that the proposed OMWS-ET is more efficient than the RRT*, in the sense that it explores many more possibilities in the vehicle/environment/task state space.

It is important to mention also, that the proposed OMWS-ET methodology is related to the adopted navigation strategy (cf. Section 5.1.2), which uses set-points based on suitable static/dynamic waypoints instead of trajectory tracking methods. The OMWS-ET method takes into account the vehicle's kinematics constraints and uncertainites as well as the used control law (cf. subsection 5.4.2). The RRT* method is more suitable for navigation strategies based on trajectory following [Karaman and Frazzoli, 2011].

	$length[\text{m}]$	$d_{border}[\text{m}]$
Voronoï	86.00	69.2931
RRT*	83.42	62.1736
OMWS-ET	82.50	65.5926

TABLE 5.1   Comparison between Voronoï, RRT* and OMWS-ET, where $length$: path length and d_{border}: minimal distance w.r.t. the nearest obstacle/limit of the way.

5.4.4.2   Local replanning for unexpected obstacles

The proposed method OMWS-ET is adapted to local replanning when an unexpected static obstacle is detected in the environment. Figure 5.21 shows the used architecture to activate the replanning of the vehicle's movements based on an initial set of waypoints, already obtained using OMWS-ET. The vehicle uses a range sensor to detect any unforeseen obstacle (cf. Figure 5.22(a)). A local replanning is activated when an obstacle is detected. This replanning takes into account the current environment state, the current vehicle pose and the current assigned waypoint to obtain a new local set of waypoints (cf. Figure 5.22(b)). If the current waypoint is unreachable (due to the obstacle configuration for instance) then the final position is replaced by the next waypoint in the list and so on. If no solution is found, then the vehicle will stop in its current pose. Figure 5.22(b) shows an example of the local replanning using an already obtained set of waypoints given in Figure 5.19(b). Finally, the vehicle moves through the new set of waypoints while guaranteeing safe navigation (cf. Figure 5.22(c)).

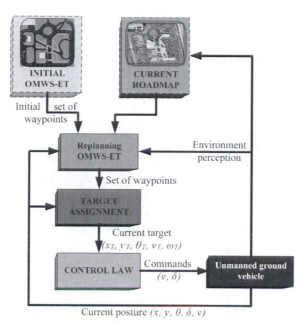

FIGURE 5.21   Schema of the local replanning.

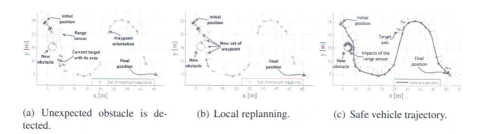

(a) Unexpected obstacle is detected.

(b) Local replanning.

(c) Safe vehicle trajectory.

FIGURE 5.22   Local replanning for an unexpected obstacle.

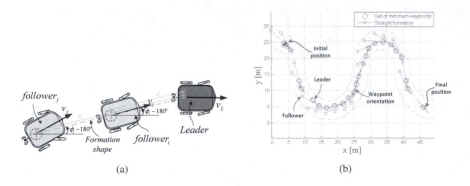

(a)   (b)

FIGURE 5.23 (a) Multi-vehicle formation (straight line shape). (b) minimum set of waypoints for multi-vehicle formation obtained by OMWS-Expanding Tree (cf. Algorithm 8).

5.4.4.3 Extension to multi-vehicle formation

The most important ideas developed in this manuscript aim to be easily extended to an even more complex system (bottom-up approach), such as multi-vehicle navigation (cf. chapter 6). The method based on expanding tree (Algorithm 8) has been extended to multi-vehicle formation where the formation is defined only according to the leader configuration [Consolini et al., 2008]. As mentioned before, the OMWS-ET algorithm takes into account the vehicle model. To deal with this multi-vehicle task, it is sufficient to adapt the term $\Delta \bar{e}_{l_{ij}}$ (cf. equation 5.18) in order to consider all trajectories of the group of vehicles. Figure 5.23(b) shows the minimum set of waypoints for a line formation ($d_i = 6 \ m$ and $\phi_i = 180°$, cf. Figure 5.23(a)) with two vehicles. The constant values are the same as the last simulations. The set of waypoints for the leader vehicle are close to the curve road boundaries because the formation needs enough space to turn while keeping the rigid formation shape. The follower (blue square) is always inside of the road boundaries.

5.5 EXPERIMENTAL VALIDATIONS

This section presents several experiments to demonstrate the efficiency of the proposal to obtain reliable, safe and flexible vehicle navigation in a structured environment. Subsection 5.5.1 will focus on the planning aspects using OMWS-ET and subsection 5.5.2 will give a complete experiment of a group of vehicles navigating in formation in an urban environment containing an obstructing obstacle. The experiments were done using VipaLab vehicles in a PAVIN platform (*Plate-forme d'Auvergne pour Véhicules INtelligents*) (cf. Annex A, page 185). A metric map of PAVIN [IP.Data.Sets, 2015] is used to perform the proposed OMWS-ET (cf. Algorithm 8). This map allows implementing the navigation through successive waypoints in a real situation. OMWS-ET computes the set of geo-referenced waypoints with optimal configurations. Some areas are restricted to guide Algorithm 8 through the PAVIN platform which has intersections and a roundabout (cf. Figure 5.24). In the presented case, these restricted areas were defined by the user.

5.5.1 Waypoints planning

OMWS-ET is used in what follows to make a comparison between two cases: the first corresponds to giving more priority to the safety criteria in equation 5.9 and the second gives more priority to the minimum angle steering rate. The analysis of the obtained solutions will be given in what follows. Moreover, the actual vehicle's trajectories are compared for these different sets of waypoints. These experiments can be found online.[1]

Figure 5.24 and 5.25 show respectively the minimum obtained set of waypoints and the corresponding vehicle's trajectories (in simulation and actual experiment).

---

[1] http://maccs.univ-bpclermont.fr/uploads/Profiles/VilcaJM/OMWS.mp4

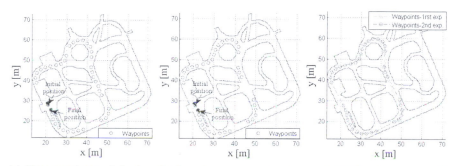

(a) First experiment: Safe planning. (b) Second experiment: Steering angle minimization. (c) Comparison between experiments.

FIGURE 5.24 Different set of obtained waypoints.

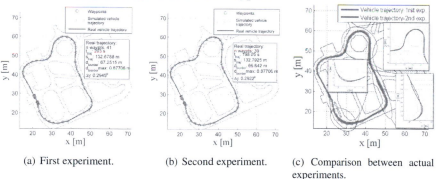

(a) First experiment. (b) Second experiment. (c) Comparison between actual experiments.

FIGURE 5.25 Actual vehicle's trajectories for different obtained sets of waypoints.

Figure 5.24(a) shows the set of waypoints of the first experiment where the constant values of the cost function (cf. equation 5.9) are $k_1 = 0.6$, $k_2 = 0.2$, $k_3 = 0.1$, $k_4 = 0.1$ and $k_h = 0.4$. The safety (k_1) has the highest priority in this experiment. Therefore, these waypoints guide the vehicle close to the middle of the route (cf. Figure 5.25(a)). Figure 5.24(b) shows the set of waypoints of the second experiment where the constant values are $k_1 = 0.3$, $k_2 = 0.2$, $k_3 = 0.4$, $k_4 = 0.1$ and $k_h = 0.4$. The minimal steering angle rate k_3 has the highest priority in this experiment. The obtained result shows that the obtained waypoints are localized very close to the road border (cf. Figure 5.25(b)).

Figure 5.24(c) and 5.25(c) show the comparison between the sets of waypoints and the real trajectories of both experiments. The velocities and steering angle of the vehicle while tracking each waypoint are shown in Figure 5.26. This figure shows the values with noise due to the encoder inaccuracies. Figure 5.25(a) and 5.25(b) show the simulated and the actual vehicle trajectories. It can be observed that they are very close (maximal error between them is less than $0.15\ m$). We can conclude therefore that the proposed optimal multi-criteria waypoint selection based on Expanding Tree (OMWS-ET, performed off-line (cf. Section 5.4.2)) permits us to deal accurately with the actual environment and experiments.

Table 5.2 shows different performance criteria to compare the set of waypoints where n_w is the number of waypoints, T is the navigation time, l_{UGV} is the traveled distance, d_{border} is the sum of minimum distance to the road boundaries and $\Delta\gamma$ is the root mean square (rms) of the steering angle rate. It is noted that the first experiment has n_w greater than the second experiment. It is due to the fact that the first experiment has safety as a priority. The proposed Algorithm 8 thus selects more waypoints to allow the vehicle to navigate as far as possible from the road borders. It

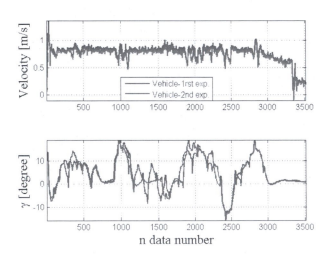

FIGURE 5.26 Vehicle velocities and steering angles evolution for each set of obtained waypoints.

can be noticed that d_{border} values are bigger in the first experiment than the second. Furthermore, the values of $\Delta\gamma$ are less in the second experiment because the highest priority was for the steering angle rate. Therefore, the vehicle can navigate with higher velocity along the trajectory and the navigation time becomes thus smaller than the first experiment.

		n_w	$T[s]$	$l_{UGV}[m]$	$d_{border}[m]$	$\Delta\gamma[°]$
1rst	Sim.	41	200	132.81	67.35	0.3123
exp.	Real	41	203	132.68	67.25	0.2945
2nd	Sim.	39	199	133.00	66.54	0.3089
exp.	Real	39	198	132.79	66.64	0.2922

TABLE 5.2   Comparison among the set of the obtained waypoints.

5.5.2   Safe and reliable multi-vehicle navigation

The navigation strategy was also tried with a pair of real urban vehicles (cf. Figure 5.27). The scenario was built to show different situations, such as multi-vehicle navigation in formation, static and dynamic target-reaching and obstacle-avoidance situations. In this experiment, each vehicle uses a combination of RTK-GPS and gyrometer to estimate its current position and orientation at a sample time of $T_s = 0.1\ s$. The vehicles have a range sensor (LIDAR) with a maximum detected range equal to $10\ m$. These sensors provide enough accurate data w.r.t. the vehicle dynamic. Indeed, in these experiments, the vehicles move at a maximum velocity of $1.5\ m/s$ due mainly to the relatively short dimensions of the used urban PAVIN platform. Moreover, the vehicles communicate by Wi-Fi, enabling the transmission of the leader's pose data.

Experiments were carried out to show the performance of the proposed control

FIGURE 5.27 (See color insert) Some images from the performed experiment.

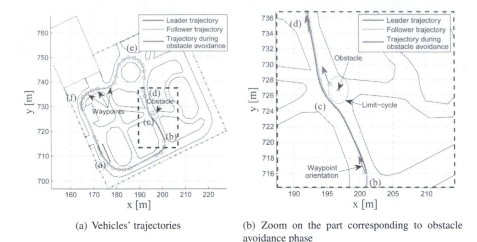

(a) Vehicles' trajectories

(b) Zoom on the part corresponding to obstacle avoidance phase

FIGURE 5.28 Vehicle trajectories obtained using GPS and a set of waypoints positioned in the environment using Algorithm 10 ($\Delta_{\alpha_{max}} = 15°$).

law and target assignment strategy using waypoint selection based on Algorithm 10 (with $\Delta_{\alpha_{max}} = 15°$) on an already defined reference trajectory. The Leader vehicle has to successively reach static waypoints. Moreover, the proposed control law (cf. section 3.2.2, page 57) was implemented in another vehicle (Follower) which takes the first vehicle (Leader) as a dynamic target to track at a curvilinear distance equal to $5\ m$ (behind the Leader). The tracking of the dynamic target allows us to apply the proposed control law to multi-vehicle systems where the dynamic set-point is given by the leader and the desired geometric formation shape [Vilca et al., 2014]. The configuration of the dynamic target is sent by the Leader to the Follower via Wi-Fi. This experiment can be found online.[2] Furthermore, to exhibit the flexibility of the proposed navigation strategy, an obstacle is placed between the waypoints. As mentioned in section 5.1.1, the proposed strategy can easily integrate the obstacle avoidance behavior (cf. Figure 5.3). Therefore, the vehicle can perform different maneuvers between waypoints, in this case obstacle avoidance, without the use of any trajectory replanning method. The used obstacle-avoidance method is based on limit-cycles as given in [Adouane et al., 2011] (cf. section 2.3, page 31). The obstacle avoidance is activated as soon as the vehicle detects at least one obstacle which can hinder the future vehicle movements toward the current assigned waypoint.

It can be seen in Figure 5.28(a) that the Leader accurately reaches the successive assigned static waypoints and the Follower also accurately tracks the dynamic target (Leader). Moreover, the Follower trajectory using the proposed control law is close to the Leader trajectory (cf. Figure 5.28(a)). Figure 5.28(b) focuses on the vehicles' trajectories when the obstacle avoidance is activated. The Leader detects the obstacle between the waypoints and it applies the reactive limit-cycle method

---

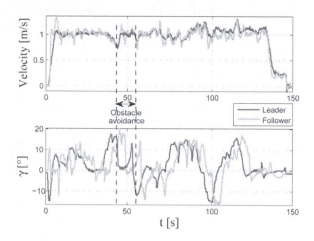

FIGURE 5.29   Control output (real experiment).

[Adouane et al., 2011] [Vilca et al., 2013a]. The Follower also avoids the obstacle since it accurately tracks the Leader trajectory. It can be noted that the proposed navigation strategy allows flexible and smooth movements between the waypoints and performance of different behaviors, such as obstacle avoidance, emergency stop or waypoint reassignment.

Figure 5.29 shows the velocity and steering angle of the vehicles. These actual values have been filtered, during the experimentation, using an Extended Kalman Filter (EKF) to reduce the sensor noise. Figure 5.30(a) and 5.30(b) show the Lyapunov function values which highlight that each vehicle control is stable and it converges

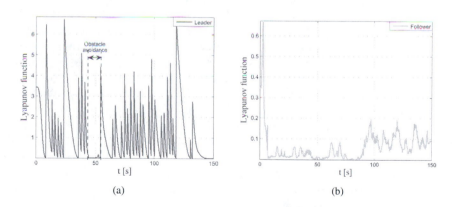

(a)   (b)

FIGURE 5.30   (a) Lyapunov function of the leader (based on static waypoint reaching) and (b) Lyapunov function of the follower (based on dynamic target tracking).

to each assigned static waypoint for the Leader and to the dynamic target for the Follower. Therefore, smooth, flexible and safe trajectories for the vehicles were obtained.

5.6 CONCLUSION

This chapter presented an overall methodology (strategy / planning / control) of autonomous vehicle navigation which uses only a set of waypoints, appropriately disposed in the environment, to perform reliable and flexible navigation in an urban environment. This methodology is an alternative to the widely used strategies of vehicle navigation which rely on the use of a pre-generated reference trajectory in order to have the vehicle's set-points. The main motivation of the proposed methodology arises from the need of further improving the navigation flexibility (to deal with different environments, tasks and contexts) while maintaining a high level of reliability and safety (cf. section 5.1).

Indeed, as shown in this chapter, the use of only a set of waypoints in the environment permits even more flexibility of the vehicle's movements, since it is allowed to perform more maneuvers between waypoints (without the necessity of replanning any reference trajectory[3]), while remaining reliable and safe (non-collision w.r.t. the road limits or any obstructing obstacle).

The proposed navigation strategy is based on successive static target reaching. The switch between targets uses, among others things, appropriate reference frames linked to the current assigned waypoint and to the next one to reach (cf. section 5.2). It has been proved in section 5.3 the reliability of the approach through the use of the asymptotic stable control law (cf. section 3.2.2, page 57) and the definition of the maximum error boundaries (distance and orientation, cf. section 5.3.1) when the vehicle is in the immediate vicinity of the current target to reach. Furthermore, the smoothness of the overall navigation has been addressed in section 5.3.2.

Once the principle of navigation strategy is validated in terms of control reliability and smoothness (cf. section 5.3), an important part of this chapter has been dedicated to the different ways to obtain the most appropriate waypoints configurations (number, poses, velocities, etc.) in the environment. Several methods to obtain these suitable sets of waypoints were presented. The main presented method, called Optimal Multi-criteria Waypoint Selection based on Expanding Tree (OMWS-ET), uses a multi-criteria function to optimize the set of obtained waypoints. This generic method takes into account the vehicle's model and uncertainties (of the environment/vehicle model) to obtain the sub-optimal set of waypoints.

A large number of simulations (cf. sections 5.3.3 and 5.4.4) and experiments (cf. section 5.5) using VIPALAB vehicles in different kinds of urban environments demonstrate the efficiency, the reliability and the flexibility of the proposed navigation strategy based on a set of discrete waypoints.

---

[3]Which could be, among other things, time-consuming and/or complex, mainly in a cluttered and dynamic environment (cf. section 5.1).

CHAPTER 6

Cooperative control of multi-robot systems

CONTENTS

T HIS CHAPTER is dedicated to the control of multi-robot systems. It constitutes a natural extension of the proposed multi-controller architectures to deal with multi-robot systems. This is possible thanks to the adopted bottom-up approach and its inherent features. The focus will be on dynamic multi-robot navigation in formation and on the cooperative strategies to perform safe, reliable and flexible navigation. An overview of other addressed multi-robot tasks (as "co-manipulation and transportation" and "exploration under uncertainty") will also be briefly presented.

6.1 INTRODUCTION (GLOBAL CONCEPTS)

6.1.1 From mono-robot to multi-robot systems

Obtaining reliable and flexible control of cooperative Multi-Robot Systems (MRS) has been and is still one of our main research objectives. The investigations made on multi-controller architectures (with their bottom-up construction (cf. section 1.4, page 15)) have been initially motivated by the control of such complex systems.

Hence, after the previous chapters dealing mainly with the control of single robots, let us exploit directly in this chapter the different developments to control an even more complex system. Obviously, several specific control mechanisms, enabling to deal with MRS, will be added. The following developments concern therefore the immersion of a mobile robot in the context of MRS. This implies among others things, that the control of this elementary robot will not depend only on its own perceptions/objectives but should also take into account the partial or overall MRS state [Adouane, 2010, chapter 1].

Controlling MRS instead of only one robot, considerably increases the control complexity, due mainly to the augmentation of the dynamic of interaction between the robots (e.g., robots can hinder each other); the number of control variables and sub-objectives to reach/achieve; the uncertainty to communicate/observe/localize the group of robots, etc. It will be shown in the following sections, how we dealt with these different aspects.

6.1.2 Cooperative robotics (definitions and objectives)

Cooperative robotics is synonymous with the existence of at least two robots which interact to perform a task. It expresses not only control of each robot individually, but also requires using an appropriate control strategy in order that the assembly of all the entities generates a coherent and efficient robots configurations to achieve the targeted tasks. The domain of cooperative robotics constitutes an active research field and is currently linked to many key application areas with great importance. Generally the use of a group of robots instead of one is motivated by two main situations: to carry out tasks which are infeasible with a single robot (e.g., moving a too heavy or bulky object) or improve certain criteria related to the rapidity, the robustness or the flexibility (redundancy of sensors and actuators provides notably better failure tolerance) of the task to achieve [Adouane, 2005]. In the latter situation, the robots join their capacities and knowledge to improve the task achievement.

As notably emphasized in sections 1.1 and 1.2 (pages 2 and 7, respectively), several current projects and important challenges are related to cooperative navi-

(a) Modular robotics and self-asembling tasks (b) Swarm robotics and heterogeneous Robot-Robot cooperation (c) Robot-Robot and Robot-Human cooperation

FIGURE 6.1 (See color insert) Different cooperative robotics projects/tasks. (a) M-TRAN project [Murata and Kurokawa, 2012], corresponds to modular robots which can autonomously reconfigure themselves to form different 2D or 3D structures. (b) Swarmanoid project; Cooperation of a swarm of ground and aerial robots to achieve complex task [Mathews et al., 2015]. (c) Object handling by multiple mobile manipulators in cooperation with a human [Hirata et al., 2007].

gation of vehicles. They are obviously not the only tasks which interest the scientific community in the large field of cooperative robotics. Figure 6.1 gives some samples of the different cooperative robotics projects/tasks which are of great interest for the robotics community. Some other examples of the use of MRS tasks can be found in exploration [Kruppa et al., 2000] [Lozenguez, 2012], transport of heavy/large objects [Alami et al., 1998] [Hichri et al., 2014a], coverage of unknown areas [Dasgupta et al., 2011] [Franco et al., 2015], etc. All these cooperative tasks find applications in different areas such as industry / service / military / agriculture / etc.

In the literature, several architectures exist to control MRS, their accurate classifications need to look over all the nuances of the used strategies: centralized *versus* decentralized (cf. section 1.3.3, page 13) and reactive *versus* cognitive (cf. section 1.3.2, page 11). In addition, if we introduce the kind of used sensors/communication, the heterogeneity or not of the MRS (e.g., in term of applied control or robots' physical structures), and so on, it appears clearly that an exhaustive classification of MRS control architecture would be a tedious work [Cao et al., 1997] [Adouane, 2005, chapters 1 to 3]. In our different proposed works in the area of MRS and even if several nuances of control architectures are addressed [Mouad et al., 2011b] [Lozenguez et al., 2013b] [Hichri et al., 2014a] [Vilca et al., 2014] [Benzerrouk et al., 2014], in our works decentralized robot coordination and local reaction to unpredictable events (i.e., without global path planning (cf. section 4.3, page 80)) are favored. The objective is to give the maximum flexibility and autonomy of the controlled MRS. It is important to notice that the navigation in formation of a group of robots is taken in this manuscript as the main targeted cooperative task. Nevertheless, an overview of other addressed multi-robot tasks will be described briefly in the following section.

6.2 OVERVIEW OF ADDRESSED MULTI-ROBOT SYSTEMS/ TASKS

In addition to the navigation in formation task, it has been investigated during our research works mainly two other cooperative tasks. These cooperative tasks use mobile robotics entities and the necessity to coordinate their movements in order to optimize the mission achievement. A summary of each work is given below.

6.2.1 Cooperative manipulation and transportation

This work aims to design and control, in the simplest way possible, cooperative robotics entities, to co-manipulate and co-transport objects of any features (shape/weight) [Adouane and Le-Fort-Piat, 2004] [Hichri et al., 2014a]. The targeted cooperative task, also called the "removal-man task" [Hirata et al., 2002], could be achieved for any payload's features in a flexible way while adjusting the number of robots and their configuration around the object. The aim is to displace it from an initial configuration to a final one in the environment. An original process of co-manipulation between robots has been proposed [Hichri et al., 2014b] [Hichri et al., 2014c]. It consists of a specific strategy which exploits the mutual robot's pushing to induce a force to lift the payload and to place it on the top of the poly-robot (the MRS) (cf. Figure 6.2(a)). Among the main challenging issues addressed in this work is to find the optimal configuration of robots around the object (cf. Figure 6.2(c)) to achieve co-manipulation and the co-transportation while maximizing the stability of the achieved task [Hichri et al., 2014a]. The stability in this work corresponds to ensure the Force Closure Grasping (FCG[1]) criterion which ensures payload stability during the co-manipulation phase and the Static Stability Margin (SSM[2]) criterion which guarantees payload stability during the co-transportation phase (cf. Figure 6.2(c)).

Several elementary navigation functions have been used to deal with this cooperative task. Among them is the "Obstacle avoidance" controller, based on limit-cycles (cf. section 2.3, page 31), which is used for two aspects: firstly when each elementary robot aims to reach its position around the payload (the robot can need to avoid other robots or any other obstacles to reach its assigned position); secondly when the overall poly-robot (the robots with the transported payload) is in the navigation phase and has to avoid any obstructing obstacle. This poly-robot navigation raises also interesting issues linked to multi-robot navigation in formation. The poly-robot is considered as an overall robot with several constraints induced by the robots' wheels composing the poly-robot [Hichri et al., 2014a] [Hichri et al., 2015]. Several works are underway to enhance the decentralized coordination of such a cooperative system.

---

[1]FCG problem is extensively studied for objects manipulation, mainly for multi-fingered robotic hand [Yoshikawa, 2010]. This problem was adapted therefore to mobile robot co-manipulation and transport.

[2]SSM are extensively studied for walking mobile robots [Wang, 2011]. This criterion has been therefore also adapted to the investigated work.

(a) From left to right: proposed sequence strategy to lift any payload (shape/weight) on the robots.

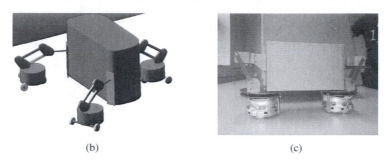

(b)   (c)

FIGURE 6.2 (See color insert) Cooperative manipulation and transportation of any payload shape/weight. (a) and (b) Multi-body dynamic simulation (with ADAMS®) for payload lifting using the developed strategy of positioning and lifting. (c) First prototype using Khepera® mobile robots and designed end-effector.

6.2.2 Cooperative exploration under uncertainty

These works deal with decentralized and cooperative exploration of different kinds of environments (open, cluttered or urban) and this, while taking into account knowledge/perception uncertainties [Lozenguez et al., 2012b]. The group of robots must coordinate, in a decentralized way, their individual explorations in order to visit all the environment while minimizing the overall robots' displacements (cf. Figure 6.3(a)). As the first step of these works, a multi-controller architecture was proposed based on a topological representation of the environment to generate a low-density map [Lozenguez et al., 2012a]. The aim is to reduce the multi-robot planning calculation costs by using only the most relevant environment's information. It is interesting to emphasize the fact that the multi-robot exploration mission has been modeled as a set of specific waypoints (locations) to visit in the environment. An optimization process was used similar to the stochastic traveling salesman problem [Kenyon and Morton, 2003].

It is to be noted that Markov Decision Processes (MDPs[3]) have been used in these works for both decision making and multi-robot coordination. More specifically a Goal Oriented MDP (GO-MDP) has been proposed to master the combinatorial explosion of such a complex system [Lozenguez et al., 2011b,

---

[3]MDP [Bellman, 1957], provides a mathematical framework for modeling decision making in situations where outcomes are partly random and partly under the control of a decision maker.

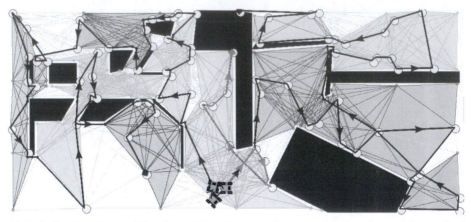

(a) Task allocation for 3 mobile robots while using appropriate waypoints definition and decentarlized exploartion policies [Lozenguez et al., 2012b].

(b)   (c)

FIGURE 6.3   (See color insert) Cooperative coordination and navigation using a group of Pioneer® in PAVIN (cf. Annex A, page 185) for (b) free area or (c) urban area [Lozenguez, 2012].

Lozenguez et al., 2013a]. The proposed solution consists of decomposing the multi-goal MDP to overcome the limitation of the number of possible considered goals [Lozenguez et al., 2011a]. Finally, successive simultaneous rounds of auctions was proposed based on individual assessments to speed up the goals/waypoints allocation [Lozenguez et al., 2013b]. Several simulations and experiments have been made to validate the different proposals in effective situations of multi-robot exploration (cf. Figures 6.3).

6.3   DYNAMIC MULTI-ROBOT NAVIGATION IN FORMATION

The focus will be made in the following sections on dynamic multi-robot navigation in formation (cf. Figure 6.4) and on the adopted strategies to perform safe, reliable and flexible navigation. In this challenging multi-robot task, the robots have to navi-

FIGURE 6.4 (See color insert) Example of autonomous navigation in formation of a group of UGVs in an urban environment (Clermont-Ferrand, France). MobiVIP project (Predit 3).

gate and keep a desired relative configuration (position and orientation) to each other or to a reference (dynamic target or trajectory). Let us present first an overview of the existing methods/strategies dealing with this kind of MRS.

6.3.1 Overview of existing strategies

The challenging navigation in formation task is one of the most important issues for MRS. In fact, many tasks require a moving MRS which must maintain a desired pattern such as in autonomous public transportation [Levinson and Thrun, 2010]; space exploration [Huntsberger et al., 2003]; military missions [Murray, 2007], agriculture [Cariou et al., 2010]; surveillance [Stoeter et al., 2002] or rescue operations [Bahr, 2009].

The literature emphasizes mainly three approaches dealing with the navigation in formation: *Virtual structure* [Mastellone et al., 2007] [Desai et al., 2001]; *Behavior-based* [Balch and Arkin, 1999] [Tang et al., 2006] and *Leader-follower* (also called the hierarchical approach in some references) [Mastellone et al., 2008] [Ghommam et al., 2010] approaches.

The virtual structure considers the formation as a single virtual body. The shape of the latter is the desired formation shape, and its motion is translated into the desired motion of each robot [Do, 2007], [Li et al., 2005]. The virtual structure is implemented in several works through potential field methods [Ogren et al., 2002], [Mastellone et al., 2007]: thus, all the members of the formation track assigned nodes which move into the desired configuration. In these works, nodes applied an attractive field to the corresponding robot, whereas obstacles and neighbor robots apply repulsive fields. Unlike motion planning, potential functions applied for the virtual structure approach use only the instantaneous and local robots' perception. The weakness of using potential functions for this approach corresponds to increasing complexity for controlling the fleet shape in a dynamic environment. In fact, it means that the robot is submitted to a frequently changing number/amplitude of forces leading to more local minima, oscillations, etc. Therefore, in this case, it is very difficult to demonstrate the robustness and the stability of the MRS navigation.

A behavior-based approach implies that each robot has a set of behavior patterns (basic tasks) to achieve. The resulting behavior of the group emerges from the basic local interaction without any explicit model of the overall cooperative behavior. However, this approach is criticized w.r.t. the way that it chooses the control for each robot. In fact, according to perception information, the control system switches between behavior patterns (e.g., competitive approach [Brooks, 1986]), or merges several controllers (e.g., motor schema [Arkin, 1989b]). This naturally makes it hard to study the stability of the global control strategy. In a distributed behavior-based approach [Antonelli et al., 2010], [Balch and Arkin, 1999], there is no hierarchy between the robots. Each one has its own perception and control [Parker, 1996].

In Leader-follower (the third approach), one or more robots are considered as Leaders, while the other robots are the followers. Generally, the leader tracks a predefined trajectory/waypoints while the followers track its transformed coordinates [Léchevin et al., 2006] [Gustavi and Hu, 2008]. Different works exploit graph theory to describe the inter-robot configuration/communication [Chen and Li, 2006, Das et al., 2002, Mesbahi and Hadaegh, 1999, Shames et al., 2011]. Several formation cases (leader reassignment, robot adding and control saturation) were presented in [Mesbahi and Hadaegh, 1999]. The authors proposed a formation control law based on the combination of Linear Matrix Inequalities and a logic based-switching system. In [Klančar et al., 2011] the followers track the leader trajectory, using platooning formation and local perceptions (camera or laser). The case of dynamic formation, i.e., the formation shape changes to another (e.g., from square to triangle), and obstacle avoidance was covered in [Chen and Li, 2006, Das et al., 2002, Shames et al., 2011]. In [Chen and Li, 2006], the leader generates a free-collision trajectory in a dynamic environment which is tracked using a formation control law based on an artificial neural network, Lyapunov function and the robot dynamic model. The stability of the dynamic formation and topology (adjacency matrix) are also demonstrated. In [Das et al., 2002], switches between different formation shapes are exploited (from triangle to line) to avoid encountered obstacles in the environment. The formation control law is based on input-output feedback linearization and on vision sensors (omnidirectional camera) embedded in each robot for localization and navigation purposes. This last approach is relatively simple to perform. However, it is noticed that a leader's failure can stop the whole system if there is no foreseen mechanism to assign another leader to the formation.

In the investigated works, the combination of the above presented approaches was exploited to enhance the overall MRS flexibility and reliability. The two main proposed control architectures are both based on a multi-controller architecture (therefore on a behavior-based approach) (cf. section 1.4, page 15) and either a *Virtual structure* (VS) approach [Benzerrouk et al., 2010b] [Benzerrouk et al., 2014] or a *Leader-follower* (LF) approach [Vilca et al., 2014] [Vilca et al., 2015a].

In what follows, in addition to presenting the proposed strategies-based VS and LF approach, several sections emphasize adopted strategies to answer mainly the following questions:

- How does the MRS ensure formation stability (reached and maintained)?

- How do the robots determine their position in the formation?

- How do the robots act if there are obstructing obstacles (static/dynamic/other robots)?

- How does the MRS adapt its formation dynamically and smoothly in order to deal with its environment dynamic and configuration?

6.3.2   Stable and reliable proposed multi-controller architecture

This section presents the proposed multi-control architecture (cf. Figure 6.5) to obtain safe and smooth robot navigation in formation. The same overall control structure, embedded in each robot, has been used for both investigated approaches: based on VS or on LF. A basic structure of type 2 (cf. Figure 2.7(b), page 39) is therefore used with notably the addition of a Formation Parameters block which determines the desired multi-robot configurations. An overview of the different blocks composing this architecture is briefly presented below while emphasizing the new blocks/features.

- "Perceptions & communication" block: As seen in the previous chapters, mainly in section 2.5.2 (page 40), this block is in charge of all the local and/or global robot perceptions. Furthermore, knowing that several robots have to coordinate their movements, it is important to have reliable and low-latency communication between either the robots themselves or with the infrastructure (cf. section 1.2, page 7).

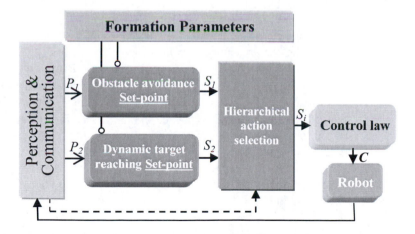

FIGURE 6.5   Multi-controller architecture embedded in each robot in the formation.

- "Hierarchical action selection" block: It aims to manage the switches between the two elementary controllers, *Obstacle avoidance* and *Dynamic target reaching* blocks, according to the formation parameters and environment perception. It activates the *Obstacle avoidance* controller as soon as it detects at least one obstacle which can hinder the robot's future movement toward its dynamic virtual target. It enables us to anticipate the activation of the obstacle avoidance controller [Adouane, 2009b] and to decrease the time to reach the assigned target (static or dynamic).

- In terms of set-point blocks, they are harmonized as motivated in section 2.4 (page 37). Indeed, they are always defined according to an appropriate target pose (x_T, y_T, θ_T) and a linear velocity v_T.

  - "Obstacle avoidance set-point" block: This well detailed block, notably in section 2.5.4.1 (page 43), permits each elementary robot to avoid reactively, and in a safe and reliable way, any obstructing obstacle. Appropriate limit-cycles are used (cf. section 2.3, page 31).

  - "Dynamic target reaching set-point" block: These set-points (cf. section 2.4.2, page 38) are defined according to the assigned formation shape (e.g., triangle, line, etc.) and the kind of used approach (VS or LF). In the VS as well as the LF approach, all the robots (except the Leader in LF approach) have to track their assigned dynamic target (given according to the desired formation).

    In the presented literature works (cf. section 6.3.1), interesting solutions for the formation control problem have been proposed, but, they are based mainly on predefined trajectories to guide the formation and to control the MRS. In the presented works, the used strategies allow us to obtain more reactive control architectures (cf. section 1.3.2, page 11) in the sense that each robot tracks the instantaneous state (pose and velocity) of its assigned virtual targets (thus, without any use of a reference trajectory or a global trajectory planning process).

    According to the use of either VS or LF approaches, some differences exist in terms of robots' set-points definition. In VS, the formation dynamic is imposed by the evolution of a virtual rigid body defined according to a global reference frame (cf. section 6.3.3). This is imposed by a central unit according to the targeted shape and the formation dynamic. Afterward, the robots have to track the assigned targets in a reactive way. In contrast, in the LF approach, the dynamic of the followers are given completely by the actual Leader's dynamic. These set-points are given therefore according to the Leader's mobile reference frame (cf. section 6.3.4).

    Sections 6.3.3 and 6.3.4 will present first the adopted formation shape modeling when respectively VS or LF approaches are used. They will also focus on the way to guarantee that the assigned target's set-points are always attainable by the robots, which permits the reliability of the overall MRS navigation in formation.

- "Control law" block: According to the used robot structure (unicycle or tricycle), a stable and generic common control law for target tracking, as given respectively in sections 3.2.1.2 (page 56) and 3.2.2 (page 57), will be used for both controllers ("Obstacle avoidance" and "Dynamic target reaching").

6.3.3 Navigation in formation using a virtual structure approach

6.3.3.1 Modeling

This strategy has been used for the navigation of a group of N Khepera® (unicycle mobile robot structure (cf. Annex A, page 185)). A set of virtual targets (points) forms a virtual structure of the same shape as the desired formation. This virtual structure is defined in [Benzerrouk et al., 2014] while following these steps:

- Define one point which is called the main dynamic target (cf. Figure 6.6),

- Define the virtual structure by using $N_T \geq N$ nodes (virtual targets) to obtain the desired geometry. Each node i is called a secondary target and is defined according to a specific distance D_i and angle Φ_i with respect to the main dynamic target. Secondary targets defined in this manner have then the same orientation θ_T as the main target. However, each target i will have its own linear velocity v_{T_i} (cf. Figure 6.7).

An example to obtain a triangular formation is given in Figure 6.6. It is clear that to have a complete distribution of control, the main dynamical target can be generated by one of the robots. This case corresponds then to the Leader-follower approach (cf. section 6.3.4).

6.3.3.2 Stable VS formation using dynamic constrained set-points

Once assigned a dynamic target for each of the robots composing the formation, these robots have to track their targets as long as there is not any obstructing obstacle (which are avoided reactively in this study). This scheme proves the stability of the overall multi-controller architecture (cf. section 3.3, page 60). However, this theoretical convergence is true only if the generated set-points are effectively attainable by the robots, considering their kinematic constraints.

To ensure the set-points attainability for the "obstacle avoidance" controller, it was proposed in [Benzerrouk et al., 2013] to constrain the robot's set-points in order to take into account the robot's nonholonomy, its maximal velocities, and the obstacles dimensions. New parameters are then introduced into the set-points formula to prevent the robot collisions. The parameter which was optimized analytically corresponds to μ (given the PELC shape (cf. Figure 2.5(a), page 34)).

For the "Dynamic target reaching set-point" block, it is given in what follows a

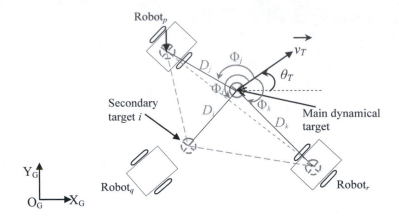

FIGURE 6.6 Maintaining a triangular formation by defining set-points according to a global reference frame (virtual structure approach).

summary of the constraints which must be imposed on the dynamic of the main target (cf. Figure 6.6) (defined by v_T and $\dot{\theta}_T$). These constraints take into account the robot's constraints and the desired formation shape. It is to be noted that to ensure reliable navigation in formation, these constraints must be verified for each elementary robot.

6.3.3.2.1 Linear velocity constraints

As given in section 3.2.1.2 (page 56), to guarantee the stability of the control law, the velocity of the target v_{T_i} to track must always verify inequality 6.1.

$$v_{T_i} \leq v_i \tag{6.1}$$

where v_i corresponds to the linear velocity control of robot$_i$ (cf. equation 3.6a, page 56) which takes into account the maximum linear robot velocity v_{max}. However, it is clear that the linear velocity of the secondary targets depends on their relative positions in the virtual structure (cf. Figure 6.7). This figure shows the different trajectories of the targets according to their relative position in the virtual structure.

The choice of D_i and angle Φ_i affects thus v_{T_i}. Each secondary target i coordinates (x_{T_i}, y_{T_i}) are expressed as

$$\begin{cases} x_{T_i} = x_T + D_i \cos(\Phi_i + \theta_T) \\ y_{T_i} = y_T + D_i \sin(\Phi_i + \theta_T). \end{cases} \tag{6.2}$$

Their derivatives are then (only rigid virtual structures are considered)

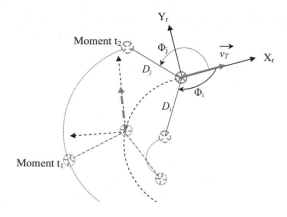

FIGURE 6.7 Virtual target trajectories to keep the virtual structure shape. Dashed curved lines represent the trajectories of the targets. Straight dashed lines illustrate the virtual structure in the previous moment.

$$\begin{cases} \dot{x}_{T_i} = \dot{x}_T - D_i \dot{\theta}_T \sin(\Phi_i + \theta_T) \\ \dot{x}_{T_i} = \dot{y}_T + D_i \dot{\theta}_T \cos(\Phi_i + \theta_T). \end{cases} \tag{6.3}$$

After several developments (detailed in [Benzerrouk et al., 2014]), it is obtained finally that the relative distance of each secondary target has to verify the following:

$$D_i < \frac{v_{max} - |v_T|}{|\dot{\theta}_T|}. \tag{6.4}$$

Note that $\left|\dot{\theta}_T\right|$ is bounded according to the developments given in the next paragraph.

6.3.3.2.2   Angular velocity constraints   The focus here is on the robot's maximum angular velocity (ω_{max}) such that the variation of the angular set-point $\dot{\theta}_T$ remains attainable. Indeed, the angular velocity applied to the robot has to verify:

$$|\omega_i| \le \omega_{max} \tag{6.5}$$

where $\omega_{max} > 0$. After several developments (detailed in [Benzerrouk et al., 2014]), it is obtained finally the permitted maximal angular velocity of these targets.

$$\left|\dot{\theta}_T\right| \le \omega_{max} - k\pi \tag{6.6}$$

where k is the constant defined in the robot's angular control law (cf. equation 3.6b, page 56). To remain attainable, the dynamic of the virtual structure has to follow two phases [Benzerrouk et al., 2014]:

1. A transitional phase, where the robots have not yet reached the formation (assigned target). In this phase, $\dot{\theta}_T$ is constrained such that $\ddot{\theta}_T = 0$.

2. Once the formation is reached, the virtual structure can vary according to equation 6.6.

6.3.3.2.3 Validation by simulation This simulation shows the importance of bounding the angular velocity of the virtual structure $\dot{\theta}_T$ according to the kinematic constraints of the robots. Hence, a mobile robot reaching a virtual target is simulated. The maximum angular velocity of the robot is set to $\omega_{max} = 3rd/s$. It is chosen $k = 0.6s^{-1}$. According to equation 6.6, and to simplify notation on figures, it is proposed to note $P = \omega_{max} - k\pi$. Based on the chosen values of ω_{max} and k, $P = 1.1$.

First, it is proposed to show the importance of the transitional phase where the variation of $\dot{\theta}_T$ must be set to 0. Hence, in Figure 6.8 (b), we can see that $\dot{\theta}_T$ increases at the beginning of the simulation (from $0.1s$) and the target trajectory follows immediately a significant curve (cf. Figure 6.8 (a)). Consequently, oscillations are observed in the robot trajectory. The robot correctly attains the target only when this one has a straight trajectory ($\dot{\theta}_T = 0$). Figure 6.8 (b) confirms this result. Naturally, the dis-

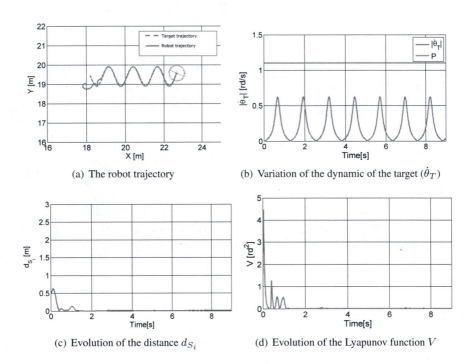

(a) The robot trajectory

(b) Variation of the dynamic of the target ($\dot{\theta}_T$)

(c) Evolution of the distance d_{S_i}

(d) Evolution of the Lyapunov function V

FIGURE 6.8 Undesirable oscillations of the robot trajectory if the transitional phase is not imposed.

tance d_{S_i} (distance robot-target) is oscillating in this case (cf. Figure 6.8.(c)). The Lyapunov function is also oscillating (cf. Figure 6.8 (d)).

Figure 6.9 shows the importance of satisfying the condition described in equation 6.6 after the transitional phase. Once the target is attained ($\dot{\theta}_T = 0$ until the moment $0.5s$), the condition 6.6 is also satisfied. It can be seen that the robot goes toward the target. Even if it increases, the variation of P is such that $\dot{\theta}_T < P$ (cf. Figure 6.9.(b)). In this interval, the robot correctly tracks its target (cf. Figure 6.9.(a)). The distance d_{S_i} separating them is $d_{S_i} = 0$ (cf. Figure 6.9.(c)). The Lyapunov function also decreases and then remains equal to 0 (cf. Figure 6.9.(d)). After $9.5s$, we remove the constraint (6.6) such that $\dot{\theta}_T$ can be $\dot{\theta}_T > P$. It can be seen that the robot cannot track the target. The oscillation of distance d_{S_i} and V confirms this (cf. Figures 6.9 (c) and (d)).

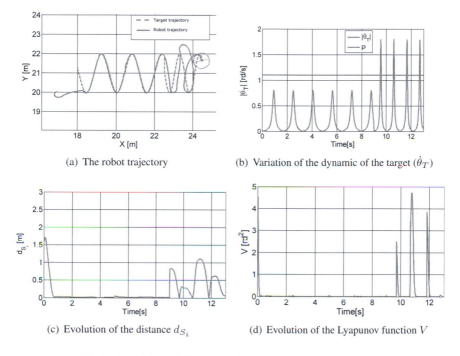

(a) The robot trajectory

(b) Variation of the dynamic of the target ($\dot{\theta}_T$)

(c) Evolution of the distance d_{S_i}

(d) Evolution of the Lyapunov function V

FIGURE 6.9 Undesirable oscillations of the robot trajectory if the imposed constraints on the dynamic target are not verified (thus if $|\dot{\theta}_T| > P$).

6.3.3.3 Safe and cooperative robot interactions

It is presented below two addressed issues linked to dynamic cooperation/coordination between robots. The proposed strategies are not only specific to unicycle robots but applicable for any mobile structure.

6.3.3.3.1 Dynamic targets allocation Each mobile robot should follow one of the secondary targets forming the geometric shape. It is interesting to optimize the allocation of the targets between the robots to rapidly reach the targeted formation shape [Ze-su et al., 2012]. Information available for each robot$_i$ are its configuration (x_i, y_i, θ_i), the one of the main virtual target (x_T, y_T, θ_T) and, D_j and $\Phi_j |_{j=1..N_T}$ (cf. Figure 6.6).

A simple idea is that each robot chooses the closest target to track. However, this may create conflicts when many robots choose the same target. To avoid this conflict, a hierarchy between them was adopted in [Benzerrouk et al., 2010b]. Hence, the desired target is given to the robot of a higher rank. However, this hierarchy does not enable us to optimize the time so that all of the formation is formed. In what follows, it is presented how each robot computes a coefficient per target to describe its interest for this one. Computed at every time interval ΔT, this coefficient informs if this target is close or far from the robot compared to the other targets. It is called *Relative Cost Coefficient* (RCC) and is noted δ. Comparing RCCs of the same target allows each robot to decide if it takes this target or gives it up to another. Hence, if this target is needed by another robot having more difficulties to find another, it is given up to this robot (a form of altruism, where the interest of the group comes before the individual interest). This joins also some works where robots imitate some human behaviors while accomplishing their tasks: for example, impatience and acquiescence [Parker, 1998], auction methods [Gerkey and Mataric', 2002] were reproduced on robots to choose their tasks. These auction methods can be divided in two different strategies: combinatorial methods which treat all possible combinations to give the optimal distribution to the MRS [Berhault et al., 2003]; repeated parallel auctions occurring every time interval to check that every robot has the suitable task [Lozenguez et al., 2013b]. The proposed RCC algorithm aims to achieve this negotiation between the robots in a simple but efficient way. The robot wins or loses a target by computing and comparing their own RCCs for these targets. Only a minimalist communication is needed between the robots.

The RCC corresponding to robot i for the target j is noted δ_{ij}. It is computed as:

$$\delta_{ij} = \frac{d_{S_{ij}}}{\sum_{k=1}^{N_T} d_{S_{ik}}} = \frac{d_{S_{ij}}}{d_{S_{ij}} + \sum_{k=1, k \neq j}^{N_T} d_{S_{ik}}} \tag{6.7}$$

where $d_{S_{ij}}$ is the distance between the robot i and the target j. For a robot i, the set of RCCs for all the targets is put in a vector Δ_i. It is clear that $0 \leq \delta_{ij} \leq 1$ (cf. equation 6.7).

Moreover, δ_{ij} is as close to 0 as

$$d_{S_{ij}} \ll \sum_{k=1, k \neq j}^{N_T} d_{S_{ik}}. \tag{6.8}$$

Thus, every robot prefers the target with the smallest RCC because it is the closest one. It is then noticed that the same result would be obtained by simply comparing the distances to the different targets and directly choosing the closest one. However, the main objective of the RCC is to negotiate the desired target with the others. Hence, if two robots i and k ask for the same target j (they are in conflict for this target), distances $d_{S_{ij}}$ and $d_{S_{kj}}$ are not sufficient to know which robot has to obtain it in order to reach faster the formation. Therefore, to negotiate their targets, robots act according to the following proposition [Benzerrouk et al., 2012a]:

Proposition *If many robots are in conflict for one target, then this target is left to the robot having the smallest RCC for this target.*

In fact, according to equation 6.8, the strategy of this proposition is to compare the situation of the robots according to the existing targets and to give up the desired one to the furthest robot from the other targets. The proposed distributed strategy for dynamic allocation of the targets is given in Algorithm 11.

Algorithm 11: Distributed virtual target assignment $(N_T \geq N)$.

Require: Vectors Δ_i, $i = 1..N$
Ensure: Choice of the virtual target to follow

  1: **while** (Target not chosen) **do**
  2:     choose the target j corresponding to the smallest RCC $\Delta_i(j)$;
  3:     **if** $\Delta_i(j) < \Delta_k(j), \forall k \neq i, k = 1..N$ **then**
  4:        go to line 11;
  5:     **else**
  6:        choose another target l such as $\Delta_i(j) < \Delta_i(l) < \Delta_i(m), \forall m \neq j$;
  7:        $j = l$;
  8:        go to line 3;
  9:     **end if**;
10: **end while**;
11: go toward the chosen target ;

The proposed algorithm is distributed on all the robots. It requires that each robot i communicates only its vector Δ_i to the other ones. It is also proposed that a vector Δ_i includes the subscript i indicating the robot identifier. Identifiers of the robots are randomly chosen and do not indicate any hierarchy for the target assignment.

According to this algorithm, every robot is able to deduce if the desired target will be really available or it will be taken by another one having a less corresponding RCC. Negotiation and allocation of the target is then done in a distributed manner. More details about RCC are given in [Benzerrouk, 2010, chapter 2]. An actual experiment using RCC is given in section 6.3.3.4.

6.3.3.3.2 Toward a null risk of collisions Given the importance of safety functions (as obstacle avoidance) for mobile robots (cf. section 2.2, page 28), mainly if the system becomes more dynamic with several interactions and reconfigurations (cf. section 6.3.4.3), this function must be addressed therefore with even more attention. Several works have been done in this area during our investigations, the most important are summarized below.

It has been favored in the presented works the use of the fewest possible perceptions/communications/negotiation to perform the proposed safety mechanisms (reactive features). In addition to static obstacle avoidance using limit-cycles (cf. section 2.3, page 31), two enhancements are emphasized according to the nature of the obstacle to avoid [Benzerrouk et al., 2013]:

1. general dynamic obstacles, and

2. robots of the same MRS, which supposes that each robot is able to recognize the other robots endowed with the same avoidance features (robots of the same homogeneous MRS).

General dynamic obstacle avoidance   Fully reactive obstacle avoidance based on PELC permits us to choose the best direction of avoidance (clockwise or counterclockwise) according to the robot configuration (cf. section 2.5.4.1, page 43). However, the corresponding decision does not take into account the dynamic of the obstacle to avoid, but only the current relative localization of the robot with regard to this obstacle (cf. Figure 2.11, page 46). First, let us recall very succinctly how we obtained the direction of reactive avoidance. The value of r (cf. equation 2.3, page 32) gives the direction of avoidance while using this simple rule:

$$r = \begin{cases} 1 & \text{if } y_{RO} \geq 0 \text{ (clockwise avoidance)} \\ -1 & \text{if } y_{RO} < 0 \text{ (counter-clockwise avoidance)}. \end{cases} \tag{6.9}$$

even if this simple rule allows us to obtain interesting results since the decision to turn clockwise or not is updated at each sample time (which permits therefore to avoid even dynamic obstacles). Nevertheless, to enhance the avoidance of dynamic obstacles it is proposed in [Benzerrouk et al., 2012b] to extend this method to dynamic obstacles without losing the control reactivity. The idea is to find the best direction of avoidance using PELC while taking into account the velocity vector of the obstacle. When a movement of the obstacle position is detected, it is considered as a dynamic obstacle by the robot. The objective for the robot is always to choose the most suitable side of avoidance (clockwise or counter-clockwise) which allows reducing the robot path to its assigned target. Nevertheless, for dynamic obstacles, the ordinate y_{RO} cannot always be used as adequate information to decide on the avoidance direction. In Figure 6.10(a), it can be noticed that if the robot decides a clockwise motion (based on its relative positive ordinate $y_{RO} \geq 0$), it fails to avoid this obstacle. In fact, the robot will go in the same direction as the obstacle (vector $\vec{v}_O$ on the figure). It may then uselessly diverge from its target by persisting in this direction.

Rather than analyzing the sign of y_{RO}, it has been proposed that the robot uses the obstacle's vector velocity $\vec{v}_O$. The idea is to project this vector on the Y_{OT} axis of the relative frame $(X_{OT}Y_{OT})$. The function r is then defined according to v_{O_y} as follows:

$$r = \begin{cases} 1 & \text{if } v_{O_y} \leq 0 \text{ (clockwise avoidance)} \\ -1 & \text{if } v_{O_y} > 0 \text{ (counterclockwise avoidance)}. \end{cases} \tag{6.10}$$

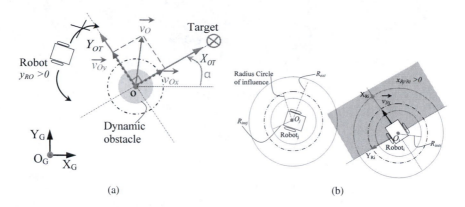

(a)   (b)

FIGURE 6.10 (a) Avoiding an obstacle. Static obstacle: the robot ordinate y_{RO} is analyzed. Dynamic one: projection of $\vec{v}_O$ is analyzed. (b) Virtual circles defining the penalty function $\psi_i^j(d_{ij})$.

By using the projection v_{O_y} of the obstacle velocity, the obstacle is always avoided round the back, such that the robot never cross the obstacle's trajectory [Benzerrouk et al., 2012b].

Obstacle avoidance between robots of the same MRS One can consider that every robot of the MRS is treated as a dynamic obstacle and projects its velocity vector to deduce the side of avoidance (cf. equation 6.10). However, some conflict problems could appear when, for instance, two robots have to avoid each other in opposite directions calculated by velocity vector projections [Benzerrouk et al., 2013]. To deal with this kind of conflict, and assuming that each robot is able to identify robots of the same system, it has been proposed to impose one reference direction for all the MRS. Hence, when one robot detects a disturbing robot of the same group, it always avoids it in a counter-clockwise direction (this can be seen as local roundabout between robots).

In addition, to avoid the inter-robot collisions it has been thus proposed a penalty function acting on the robot's linear velocities [Benzerrouk et al., 2012b]. Furthermore, this function enables us to enhance the trajectories' smoothness and avoid oscillating robots movements [Vilca et al., 2014]. Each robot is enclosed by two circles C_{int} and C_{ext} with respectively radius R_{int} and R_{ext} ($R_{int} < R_{ext}$) (cf. Figure 6.10(b)). The collision occurs when the distance d_{ij} between robot$_i$ and robot$_j$ is less than R_{int}. Hence, the penalty function ψ_i^j for the robot$_i$ w.r.t. the robot$_j$ is defined as:

$$\psi_i^j(d_{ij}) = \begin{cases} 1 & \text{if } d_{ij} \geq R_{ext} \\ (d_{ij} - R_{int_i})/(R_{ext} - R_{int_i}) & \text{if } (R_{int_i} < d_{ij} < R_{ext}) \text{ and } (x_{O_j/R_i} > 0). \\ 0 & \text{if } d_{ij} \leq R_{int_i} \end{cases}$$

$$(6.11)$$

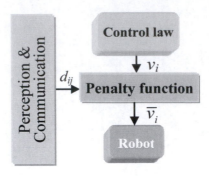

FIGURE 6.11 Integration of the penalty function in the proposed architecture.

The modified linear velocity of the robot$_i$ is then given by:

$$\bar{v}_i = v_i \psi_i^j. \tag{6.12}$$

Using the definition of R_{int_i} (where $R_{int_i} \neq R_{int_j}$), it is guaranteed that two robots do not stop simultaneously. Indeed, if the robots have the same R_{int_i} we can observe local minima in certain configurations, in fact, when $d_{ij} < R_{int_i}$ then $\psi_i^j = \psi_j^i = 0$ and the robots are stopped at the same time. R_{ext} is fixed according to communication constraints (latency) and localization errors for instance. More details about the choice of R_{int}, R_{ext} and the extension of this method for more than 2 robots is given in [Benzerrouk et al., 2012b]. This penalty function can be easily integrated in the proposed control architectures (cf. Figure 6.5), by adding a block after the output of the *Control law* block (cf. Figure 6.11).

6.3.3.4 *Experimental results*

Experiments were performed using Khepera® robots. The MRS perceptions have been centralized. Hence, navigation was achieved on a platform equipped with a camera giving positions and orientations of the robots by detecting the bar code associated with each one (cf. Annex A, page 185). This information was sent to the robots by a computer through a Wi-Fi network. In [Benzerrouk et al., 2010b], the virtual structure has a straight trajectory. It is presented in what follows to use circular motion such that all the targets remain attainable by all the robots despite their kinematic constraints [Benzerrouk et al., 2014]. Knowing that the dynamic of the virtual structure has to follow equation 6.6, the radius R_{vs} of the circular motion formed by the main target T_1 (cf. Figure 6.12(a)) verify

$$R_{vs} = \frac{v_T}{\dot{\theta}_T} > \frac{v_T}{\omega_{max} - k\pi} \tag{6.13}$$

with v_T constant and $v_T \ll v_{max}$.

First, a clockwise motion is considered (cf. Figure 6.12(a)). It is observed that the robots converge to the virtual structure even without passing the transitional phase.

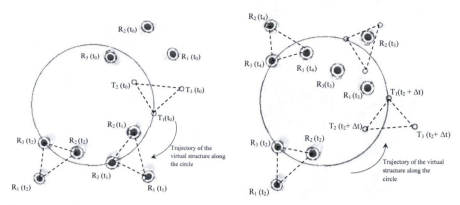

(a) $t_0 \rightarrow t_2$: clockwise motion of the virtual struc-ture

(b) $t_2 \rightarrow t_4$: switching to counter-clockwise motion

FIGURE 6.12   Real trajectory of the robots (top views from the camera). Distributed allocation (a) and reallocation (b) of the targets. Notation: $T_i(t_j)$ Target i at moment j, $R_i(t_j)$ Robot i at moment j.

The reason is that R_{vs} is big enough and initial conditions of the robots are far from the critical situations described in [Benzerrouk et al., 2014]. At moment $t_2 + \Delta t$, a jump of the virtual structure state is produced (cf. Figure 6.12(b)). Also, the dynamic of the virtual structure is changed so that its motion becomes counter-clockwise. The distances between the robots and their targets are given in Figure 6.13. They decreased to 0, which confirms that the formation was reached and maintained. When the virtual structure dynamic was changed, the robots were far from their targets, which explains the observed jumps. The same observations were noticed on the global Lyapunov function (cf. Figure 6.14).

In terms of RCC use, at the beginning of the experiment (at $t = t_0$ (cf. Figure 6.12(a))), every robot calculates the RCC for all the targets. Results are given in Table 6.1. For the robot R_1, the smallest RCC corresponds to T_3. This one is not desired by any other robot since the RCC of R_2 and R_3 for this target is not the smallest one. However, R_2 and R_3 both ask for T_2 through their RCC. Since R_2 has the smallest one, R_3 has to look for another. It takes the remaining target T_1. The second phase starts at $t_2 + \Delta t$ (cf. Figure 6.12(b)), and the robots recalculate again their RCC for each target. The RCC are given in Table 6.2. This table shows that all the robots prefer target T_2. R_1 obtains it because it has the smallest corresponding RCC. R_2 and R_3 search then for the target with the RCC immediately higher than the RCC of T_2. Again, both are interested in T_1. The latter is obtained by R_3 because its RCC is smaller. R_2 takes the remaining target T_3. It can be seen that R_2 and R_3 give up T_2 to R_1 (a kind of altruism is thus observed [Benzerrouk et al., 2012a] [Adouane and Le-Fort-Piat, 2004]).

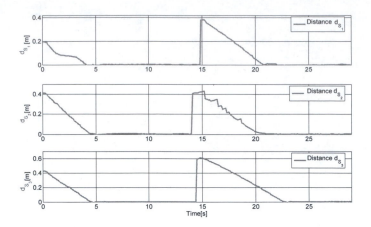

FIGURE 6.13 Variation of the distance d_{S_i} between robot i and the chosen target $(i = 1..3)$.

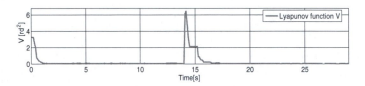

FIGURE 6.14 Evolution of the global Lyapunov function $V = \sum_{i=1}^{3} V_i$.

	T_1	T_2	T_3
R_1	0.41	0.32	0.25
R_2	0.39	0.23	0.33
R_3	0.39	0.24	0.41

	T_1	T_2	T_3
R_1	0.36	0.21	0.38
R_2	0.37	0.22	0.40
R_3	0.34	0.26	0.40

TABLE 6.1 Relative cost coefficient at moment (t_0).

TABLE 6.2 Relative cost coefficient at moment $(t_2 + \Delta t)$.

6.3.4   Navigation in formation using leader-follower approach

The navigation in formation is addressed in what follows while using the *Leader-follower* approach [Vilca et al., 2014]. This approach has been adopted and applied on VIPALAV vehicles (cf. Annex A, page 185). Several simulations and experiments will be given in the following sub-sections to validate the different proposals linked to the stability (cf. section 6.3.4.2) and the dynamic reconfiguration (cf. section 6.3.4.3) of the fleet of UGVs.

6.3.4.1 Modeling

The *Leader-follower* approach enables us to maintain a rigid geometric shape (e.g., a triangle in Figure 6.15). The formation is defined in this case w.r.t. the Cartesian frame (local frame of the leader). The proposed formation, based on *Leader-follower* approach, is defined by:

- A Leader (UGV_L in Fig. 6.15); its pose (x_L, y_L, θ_L) and its linear velocity v_L determine the dynamic of the formation.

- The formation structure is defined with as many nodes as necessary to obtain the desired formation shape. Each node i is a virtual dynamic target (T_{d_i}). The formation is defined as $\mathbf{F} = \{\mathbf{f}_i, i = 1 \cdots N\}$, where $\mathbf{f}_i$ are the coordinates $(h_i, l_i)^T$ of the dynamic target T_{d_i} w.r.t. the leader local reference frame.

The position and orientation of each node (virtual target) are computed from the leader configuration. The leader position determines the node positions according to the formation shape. The instantaneous center of curvature I_{cc_L} of the formation is determined by the leader according to its movements (cf. Figure 6.15). I_{cc_L} allows us to compute the desired orientation of the nodes according to the formation shape.

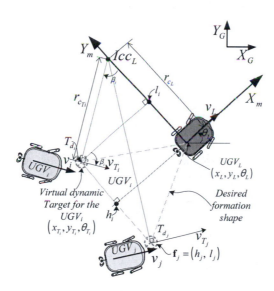

FIGURE 6.15 Maintaining a triangular formation by defining set-points according to a mobile reference frame linked to the Leader (Leader-follower approach).

The leader turns around I_{cc_L} (positioned perpendicularly to its rear wheels), then the other target set-points T_{d_i} must also turn around I_{cc_L} to maintain a rigid formation. Thus, the target velocity v_{T_i} must be tangent to the circle which has I_{cc_L} as center and the distance between T_{d_i} and I_{cc_L} as radius r_{cT_i}.

The idea behind this strategy is to eliminate the dependency of each UGV to a global reference frame. A straightforward transformation can be applied to obtain the set-point w.r.t. a local reference frame attached to the leader. The polar coordinates (r_i, Φ_i) can also be used by applying a straightforward transformation. An important advantage of the used *Leader-follower* approach is that it does not depend here on any reference trajectory and the formation is fully defined by the instantaneous dynamic of the leader. Furthermore, the presented approach is more reactive in the sense that it takes at each sample time only the current configuration and velocity of the Leader, instead of using the trajectory of the Leader as a reference for the formation [Chen and Li, 2006, Shames et al., 2011].

An important consideration to take into account to achieve the presented formation strategy, is that the followers have to know, as accurately as possible, the leader state (pose and velocity). It is assumed in what follows that the leader sends its state by stable Wi-Fi communication without latency. However, cameras and/or LIDAR sensors embedded in each follower, can be used to estimate the leader state [Das et al., 2002] [El-Zaher et al., 2012] [Vilca et al., 2015a].

In the sequel, $\mathbf{f}_i$ is given in a global Cartesian frame to homogenize the notation of the equations. The pose of the virtual target T_{d_i} w.r.t. the leader pose in the global reference frame can be written as (cf. Figure 6.15):

$$
\begin{cases}
x_{T_i} &= x_L + h_i \cos(\theta_L) - l_i \sin(\theta_L) \\
y_{T_i} &= y_L + h_i \sin(\theta_L) + l_i \cos(\theta_L) \\
\theta_{T_i} &= \theta_L + \beta_i
\end{cases}
\tag{6.14}
$$

where (x_L, y_L, θ_L) is the current pose of the leader and β_i is the T_{d_i} orientation w.r.t. the leader pose. It is given by:

$$
\beta_i = \arctan\left(h_i/(r_{c_L} - l_i)\right)
\tag{6.15}
$$

where r_{c_L} is the radius of curvature of the leader. Differentiating equation 6.14, the velocities of each T_{d_i} are given thus by:

$$
v_{T_i} = \sqrt{(v_L - l_i \omega_L)^2 + (h_i \omega_L)^2}
\tag{6.16}
$$

$$
\omega_{T_i} = \omega_L + \dot{\beta}_i
\tag{6.17}
$$

where v_L and ω_L are respectively the linear and angular velocities of the leader, $\dot{\beta}_i$ is computed as:

$$
\dot{\beta}_i = -h_i \dot{r}_{c_L} / \left((r_{c_L} - l_i)^2 + (h_i)^2\right).
\tag{6.18}
$$

One can note from equation 6.18 that when $\dot{\beta}_i$ is equal to zero, the formation has a constant radius of curvature r_{c_L} and the angular velocities of the virtual targets are equal to the angular velocity of the leader ($\omega_{T_i} = \omega_L$) (cf. equation 6.17).

6.3.4.2 Stable LF formation using dynamic constrained set-points

This part presents an adaptive computation of the leader constraints (its maximum linear velocity and steering angle) to obtain the dynamic of each virtual target (thus the set-points for the followers) which satisfy the vehicles constraints of the over-all formation [Vilca Ventura, 2015]. These adaptive constraints allow improving the rapidity and the steady convergence of the desired formation shape.

The local frame of the leader allows us to keep a constant geometric structure during the navigation of the group of UGVs (cf. Figure 6.15). The dynamic of this geometric structure is subordinated to the dynamic of the leader (cf. equations 6.14, 6.16 and 6.17). Furthermore, the dynamic of the virtual targets (cf. Figure 6.15) cannot be greater than a maximum values which must be verified to reach and maintain the formation.

It is considered in what follows a homogeneous vehicles system, i.e., all UGVs have the same physical constraints. The UGVs are modeled as a tricycle, then the linear velocity, steering angle and acceleration of the followers are constrained by v_{min}, v_{max}, γ_{max} and a_{max}. Hence, they must satisfy:

$$v_{min} \leq |v_{T_i}| \leq v_{max} \tag{6.19}$$

$$|\gamma_{T_i}| \leq \gamma_{max} \tag{6.20}$$

$$|\dot{v}_{T_i}| \leq a_{max} \tag{6.21}$$

The dynamic of the virtual targets according to the leader dynamic are given using equations 6.16 and 6.17. Therefore, the leader constraints such as velocity, steering angle and acceleration can be defined as functions of the UGV constraints (cf. equations 6.19, 6.20 and (6.21) and the formation shape.

The steering angle is directly related to the curvature of the vehicle $c_c = 1/r_c = tan(\gamma)/l_b$. Therefore, the steering angle constraint (cf. equation 6.20) can be written as a curvature constraint. Moreover, $\omega_L = v_L c_{c_L}$. This representation as a function of c_{c_L} is useful for the following computation. To simplify the notation, let us introduce:

$$A(c_{c_L}) = A_{c_L} = (1 - l_i c_{c_L})^2 + (h_i c_{c_L})^2. \tag{6.22}$$

By introducing (6.16) and (6.22) in the velocity constraint (6.19), it is obtained:

$$v_{min} \leq v_L A_{c_L}^{1/2} \leq v_{max}. \tag{6.23}$$

The steering angle constraint can be written as a function of curvature $c_{c_{max}}$. Using (6.16) and (6.17) to compute the curvature of the followers $c_{c_{T_i}} = \omega_{T_i}/v_{T_i}$ with (6.18), it is obtained:

$$\left| \frac{c_{c_L}}{A_{c_L}^{1/2}} + \frac{h_i \dot{c}_{c_L}}{v_L A_{c_L}^{3/2}} \right| \leq c_{c_{max}} \tag{6.24}$$

where $|\cdot|$ is the absolute value of the expression.

The acceleration constraint (6.21) is obtained by deriving v_{T_i} (6.16):

$$\left| v_L A_{c_L}^{-1/2}((l_i^2 + h_i^2)c_{c_l} - l_i)\dot{c}_{c_L} + \dot{v}_L A_{c_L}^{1/2} \right| \leq a_{max}. \qquad (6.25)$$

An important term for the constraints is $\dot{c}_{c_L}$. For the tricycle, it is related to the velocity of the steering angle of the leader which is given by:

$$\dot{c}_{c_L} = \dot{\gamma}_L \sec^2(\gamma_L)/l_b \qquad (6.26)$$

where γ_L is the steering angle of the leader and sec is the secant function.

Analyzing (6.23), (6.24) and (6.25), we note that the limits of (6.22) allow us to obtain the leader constraints. The first derivate of $A(c_{c_L})$ is computed to obtain the minimum value of (6.22):

$$h_m^2/\left(h_m^2 + l_m^2\right) = A_{c_L min} \leq A(c_{c_L}) \leq A(c_{c_L max}) \qquad (6.27)$$

where $\mathbf{f}_m = (h_m, l_m)^T$ are the coordinates of the farthest node w.r.t. the instantaneous center of rotation determined by the leader I_{cc_L}. The limit of $\dot{c}_{c_L}$ is given by:

$$\dot{c}_{c_L} \leq \dot{c}_{c_L max} = \dot{\gamma}_{Lmax} \sec^2(\gamma_{Lmax})/l_b \qquad (6.28)$$

where γ_{Lmax} is the maximum steering angle of the leader.

Using equations 6.27 and 6.28 and applying the triangle inequality in equations 6.23, 6.24 and 6.25, it is obtained:

$$v_{Lmax} A^{1/2}(c_{c_L max}) < v_{max} \qquad (6.29)$$

$$\left| \frac{c_{c_L max}}{A_{c_L min}^{1/2}} \right| + \left| \frac{h_m \dot{c}_{c_L max}}{v_{Lmin} A_{c_L min}^{3/2}} \right| < c_{c max} \qquad (6.30)$$

$$\left| \frac{v_{Lmax}}{A_{c_L min}^{1/2}}((l_m^2 + h_m^2)c_{c_L max} - l_i)\dot{c}_{c_L max} \right| + \left| a_{Lmax} A_{c_L max}^{1/2} \right| < a_{max} \qquad (6.31)$$

where $v_{Lmin} = v_{min}/A_{c_L min}^{1/2}$.

Finally, we obtain the leader constraints v_{Lmax}, $c_{c_L max}$ and a_{Lmax} which respect all the physical vehicle (followers) constraints while solving numerically the three inequalities given in equations 6.29, 6.30 and 6.31. Nevertheless, these fixed leader constraints can reduce drastically the dynamic of the leader (velocity, steering angle and acceleration close to their minimum values) and therefore the dynamic of the formation. To achieve this, we proposed an adaptive constraints, velocity and steering angle using the dynamic of the leader, which permits to improve the convergence toward the desired formation and to keep it. The proposed adaptive constraints of the Leader are given by:

$$v_{Lmin} = v_L \tag{6.32}$$

$$v_{Lmax} = v_{max} A_{c_L}^{-1/2} \tag{6.33}$$

$$\left| \frac{c_{c_L max}}{A_{c_L min}^{1/2}} \right| = c_{c_{max}} - \left| \frac{h_m \dot{c}_{c_L}}{v_{Lmin} A_{c_L}^{3/2}} \right| \tag{6.34}$$

where $A(c_{c_L})$ and $\dot{c}_{c_L}$ are the instantaneous values according to the current c_{c_L} of the leader. These adaptive constraints are obtained from equations 6.29 and 6.30 to get the maximum values (velocity and steering angle) according to the leader curvature (cf. equation 6.22). The adaptation of v_{Lmin} and the instantaneous values of $\dot{c}_{c_L}$ allow us to increase the limits of γ_{Lmax} while decreasing the second term of equation 6.30. Furthermore, c_{c_L} contributes to increase the limits of v_{Lmax}, e.g., when the vehicle is moving in a straight line, the formation can navigate with v_{max}.

Simulation results   This section shows the evolution of a group of $N = 3$ UGVs navigating in a cluttered environment while keeping the desired LF formation. All simulations were made in MATLAB® software. The physical parameters of the used UGV are based on the urban vehicle VIPALAB (cf. Annex A, page 185). The UGV constraints are minimum velocity $v_{min} = 0.1\ m/s$, maximum velocity $v_{max} = 1.5\ m/s$, maximum steering angle $\gamma_{max} = \pm 30°$ and maximum acceleration $a_{max} = 1.0\ m/s^2$. It is considered that the sample time is $0.01\ s$. Each UGV has a range sensor (LIDAR) with a maximum detected range equal to $D_{max} = 10\ m$ and a stable communication network. The controller parameters are set to $\mathbf{K} = (1,\ 2.2,\ 8,\ 0.1,\ 0.01,\ 0.6)$ (cf. section 3.2.2, page 57). These parameters were chosen to obtain a safe and smooth trajectory, fast response and velocity value within the limits of the vehicle capacities.

For each simulation the vehicles start at the same configuration and must reach the same final configuration. One UGV is considered as the leader, i.e., the formation will be defined according to its configuration. We consider a triangle shape $\mathbf{F} = ((-4, -2)^T, (-4, 2)^T)\ m$ (cf. Figure 6.15). The initial positions of the vehicles have an offset $(\Delta x, \Delta y) = (1,\ 0.5)\ m$ from the initial position of their assigned virtual targets. The group of UGVs has to keep the formation while moving in a cluttered environment. A static target is defined in the environment, and the leader (and thus the formation) has to go toward it while avoiding any disturbing obstacle.

The group of UGVs navigates in triangular formation. When the leader detects an obstacle with adequate range, then the whole formation will avoid it while keeping the desired shape using the limit-cycle method. K_p corresponding to a PELC safe margin (cf. equation 2.3, page 32) is increased by $2\ m$ to allow a safe formation navigation. Indeed, limit-cycles obstacle avoidance navigation has been proposed initially for the navigation by single robots, and it is extended in this work to the case of a group of UGVs. Knowing that the formation is defined by longitudinal h_i and lateral l_i coordinates (cf. equation 6.14) (cf. Figure 6.15), the value of K_p is chosen

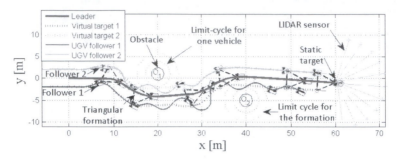

(a) Without adaptive leader constraints

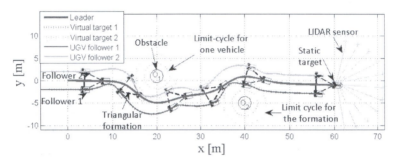

(b) With adaptive leader constraints

FIGURE 6.16 Navigation in formation $\mathbf{F}$ of a group of $N = 3$ UGVs.

in order to take into account the maximum lateral coordinate of the formation shape l_{imax}. The advantage of the proposed method is to maintain the shape of the whole formation even when obstacles hinder the formation navigation, instead of having each robot locally avoid the obstacles [Benzerrouk et al., 2012a].

The followers track their virtual target to keep the desired formation $\mathbf{F}$. Figures 6.16(a) and 6.16(b) show respectively the trajectories of the UGVs without leader constraints and using the proposed adaptive leader constraints. Figures 6.17(a) and 6.17(b) show respectively the velocities and steering angles of the UGVs without adaptive leader constraints and using the proposed adaptive constraints. The presented method allows us to obtain smooth (values of the vehicle commands) and safe (obstacle avoidance) navigation in formation (cf. Figures 6.16(b) and 6.17(b)). The trajectories of the formation without leader constraints have some oscillations which are related to the case when the velocities of the virtual targets are greater than v_{max} (cf. Figure 6.17(a)). Furthermore, it is noted in Figure 6.17(b) that the velocities and steering angle of the followers comply with their physical constraints when the proposed adaptive leader constraints are applied.

Figure 6.18 shows the values of errors d and e_θ between each UGV and its virtual target. For the formation using adaptive leader constraints, it is observed in Figure 6.18 some small peaks that are related to the fast dynamic change of the leader (dy-

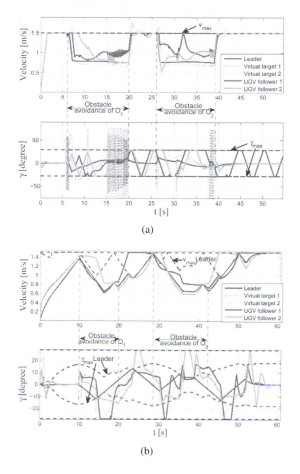

FIGURE 6.17 Commands of the UGVs (a) without dynamic leader constraints, and (b) while using the presented adaptive leader constraints.

namic of the formation is increased and the saturation occurs in the followers (cf. Figure 6.17(b))) when the leader curvature is incremented.

To quantify the distortion between the desired formation **F** and the real obtained formation shape, the Procrustes shape distance was used [Kendall, 1989] [Ze-su et al., 2012]. Basically, the Procrustes distance P_d is a least-squares type shape metric that requires aligned shapes with one-to-one point correspondence. Figure 6.19 shows the evolution of the Procrustes distance P_d between the positions of the vertex for different leader velocities. It is observed that the formation without leader's constraints does not converge to the desired formation shape **F**. It occurs when the dynamic constraints are not satisfied by the virtual targets. The followers cannot reach their assigned targets and the convergence of the formation is therefore not guaranteed (cf. Figures 6.18 and 6.19, dotted lines). Contrary to this, the formation using the proposed adaptive leader's constraints converges to the desired formation shape **F** (cf. Figures 6.18 and 6.19, continuous lines).

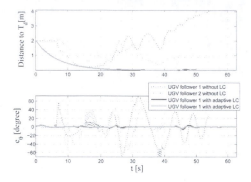

FIGURE 6.18 Distance and orientation errors of the UGVs w.r.t. their virtual targets with and without adaptive Leader Constraints (LC).

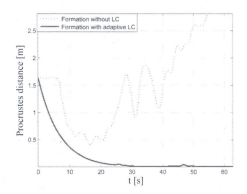

FIGURE 6.19 Formation shape **F** during the navigation with and without adaptive Leader Constraints (LC).

6.3.4.3 Dynamic and smooth formation reconfiguration

The challenge consists in what follows to guarantee the stability and the safety of the multi-vehicle system at the time of transitions between configurations (e.g., line toward square, triangle toward line, etc.) [Vilca et al., 2014]. This will make it possible to change online the formation shape of the MRS according to the context of navigation (e.g., to pass from a triangle configuration toward a line if the width of the navigation way is not sufficient).

Different methods dealing with formation reconfiguration have been proposed in the literature [Chao et al., 2012, Chen and Li, 2006, Das et al., 2002]. Many methods exploit Model Predictive Control (MPC) based on time horizon and optimization

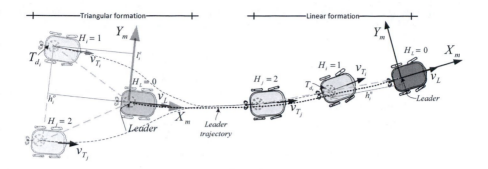

FIGURE 6.20   Formation reconfiguration between, for instance, triangular and linear formation shapes.

of a cost function [Chao et al., 2012]. These methods are generally time-consuming due to predictive computation w.r.t. a time horizon. Moreover, they were applied to small unicycle robots and are based on predefined trajectories computed along this time horizon. It is presented in what follows a new Strategy for Formation Reconfiguration (SFR) based on suitable smooth switches between different virtual target configurations.

It is considered in the following a deterministic target assignment, and a label H_i of the virtual target T_{d_i} is assigned to UGV$_i$ at the beginning of the experiments. This label is kept by each UGV in the reconfiguration phase. It is important to notice that the new virtual targets (defined on the new formation shape) must be ahead of the UGVs to guarantee the stability of the overall system (the vehicle must not go back to reach the new virtual target). If this condition is not satisfied, then the former formation will be adapted by increasing smoothly and contentiously the longitudinal coordinates h_i until all UGVs are positioned in the right configuration. The error between the coordinates of the former and the new formation $\mathbf{e}_{f_i}(e_{h_i}, e_{l_i})$ is defined as:

$$\mathbf{e}_{f_i} = \mathbf{f}_i^n - \mathbf{f}_i^f \qquad (6.35)$$

where $\mathbf{f}_i^f(h_i^f, l_i^f)$ and $\mathbf{f}_i^n(h_i^n, l_i^n)$ are respectively the coordinates of the former formation and the new desired formation (cf. Figures 6.15 and 6.20).

The reconfiguration process between the different formation shapes is given by:

$$\mathbf{f}_i = \begin{cases} h_i = h_i^n - e_{h_i}e^{-k_r(t-t_r)}, & l_i = l_i^n; \quad \text{if } e_{h_i} < 0 \\ h_i = h_i^n, & l_i = l_i^n; \quad \text{if } e_{h_i} \geq 0 \end{cases} \qquad (6.36)$$

where $\mathbf{f}_i(h_i, l_i)$ are the coordinates of the current virtual target T_{d_i} to be tracked by the follower UGV$_i$. e_{h_i} is the longitudinal coordinate of $\mathbf{e}_{f_i}$ that enables to detect if the virtual target is ahead of its corresponding follower ($e_{h_i} \geq 0$). The adaptation function when $e_{h_i} < 0$ (virtual target behind to follower$_i$) is set as proportional to the error between formation shapes, where k_r is a real positive constant designed according to the dynamic of the leader and $t_r > 0$ is the initial time for the reconfiguration process.

6.3.4.4 Simulations and experiments

This section shows the navigation of a group of $N = 3$ UGVs in a cluttered environment. The reconfiguration strategy (SFR) between the formation shapes is analyzed. The radius for non-collision between UGVs (cf. section 6.3.3.3.2) are selected as $R_{int_L} = 1.8\ m$, $R_{int_1} = 2.2\ m$, $R_{int_2} = 2.0\ m$ and $R_{ext} = 2.7\ m$. The UGVs' control parameters and features are those given in section 6.3.4.3. For each simulation the vehicles start at the same configuration and must reach the same final configuration. The initial positions of the vehicles have an offset $(\Delta x, \Delta y) = (1,\ 0.5)\ m$ from the initial position of their assigned virtual targets.

Figure 6.21 shows the navigation of 3 UGVs in a cluttered environment. It is considered that the initial formation coordinates are defined by $\mathbf{F} = (\mathbf{f}_1, \mathbf{f}_2)$, with $\mathbf{f}_1 = (-4, -2)^T\ m$ and $\mathbf{f}_2 = (-4, 2)^T\ m$ (triangular shape). Therefore, the group of UGVs must keep the formation while moving in a cluttered environment. A static target is defined in the environment, the leader (and thus the formation) must go toward it while avoiding the hindering obstacle. The new targeted formation is defined as a straight line with the following coordinates $\mathbf{F}^n = (\mathbf{f}_1^n, \mathbf{f}_2^n)$, with $\mathbf{f}_1^n = (-6, 0)^T\ m$ and $\mathbf{f}_2^n = (-3, 0)^T\ m$.

At the beginning of the simulation (cf. Figure 6.21), the navigation of the group of UGVs is in triangular formation $\mathbf{F}$. When the leader detects an obstacle with adequate range to allow the formation reconfiguration, then the leader avoids the obstacle using the limit-cycle method [Adouane et al., 2011] and sends the new desired formation $\mathbf{F}^n$ to the other UGVs (followers) to reconfigure the formation. The formation returns to triangular shape $\mathbf{F}$, when the leader does not detect obstacles that can hinder the other UGVs' movements and the last follower left behind the avoided obstacle. The

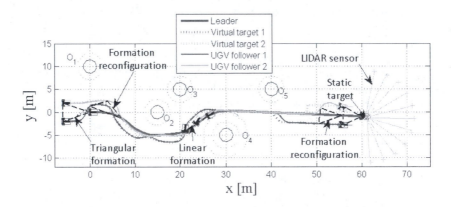

FIGURE 6.21 Navigation with reconfiguration in formation for a group of $N = 3$ UGVs.

adaptation phase enables to have the virtual target always ahead of the followers to obtain a suitable adaptive formation reconfiguration (cf. Figures 6.22(a) to 6.23(b)).

Figure 6.22(a) shows the values of errors d and e_θ between each UGV and its virtual target. At first reconfiguration, it can be observed that the follower 1 waits until its assigned virtual target is ahead. Moreover, it is noted some small peaks that are related to the fast dynamic change of the leader (the dynamic of the formation increased and the saturation occurs in the followers when the leader curvature is increased). Figure 6.22(b) shows the distances between the UGVs. This last figure shows clearly the non-collision between the UGVs in the formation, i.e., $d_{ij} > R_{int_{12}}$.

Figures 6.21 and 6.23(a) show respectively the trajectories and the velocities of the UGVs. It can be noted that the vehicles' trajectories are smooth along the navigation and there are neither collisions with the obstacles nor inter-vehicle collisions. The reconfiguration strategy was designed to reduce the peaks of the control commands of each UGV when the transition between the formations occurs (cf. Figure

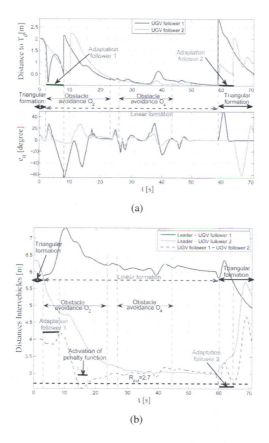

FIGURE 6.22 (a) Distance and orientation errors of the UGVs w.r.t. their virtual targets. (b) Distance among the UGVs.

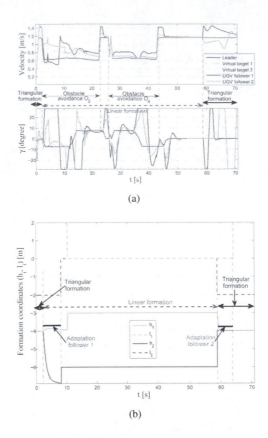

FIGURE 6.23 (a) Velocities commands of the UGVs. (b) Progress of the set-points definition $\mathbf{f}_i$ according to the proposed SFR.

6.23(a)). The proposed strategy allows thus to adapt the formation according to the environment context. Figure 6.23(b) shows the evolution of the formation coordinates (h_i, l_i) (virtual target positions). It can be observed that the adaptation phase of h_i when the follower is always ahead of its new assigned virtual target (cf. equation 6.36) which attests to the efficiency of the presented reconfiguration strategy.

An experimentation was made using 3 VIPALABs. The objective is to validate the proposed strategies based on the Leader-follower approach and reconfiguration mechanism. Figure 6.24(a) shows the sequence of the MRS evolution, from the beginning with the initial triangular formation to a linear one, when the Leader detects an obstacle, and once the last follower detects the end of the obstacle, the formation returns to the triangular formation in a smooth way. Figure 6.24(b) shows the trajectories of the 3 VIPALABs, and it attests to the safety and the smoothness of the performed navigation in formation and its reconfiguration.

Final demonstration - SafePlatoon project 26 September 2014

(a)

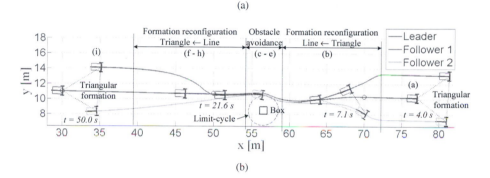

(b)

FIGURE 6.24 (See color insert) Final demonstration given in the context of the SafePlatoon project.

6.4 CONCLUSION

This chapter has focused on the control of multi-robot systems. This control was made relatively easier thanks to the use of a bottom-up approach, followed since the beginning of this manuscript. It constitutes a natural extension of the previous presented multi-controller architectures in order to deal with complex multi-robot systems.

After giving an introduction on the general definitions / concepts / tasks / potentialities linked to Multi-Robot Systems (MRS), an overview was presented of our different addressed multi-robot tasks. Indeed, in addition to the navigation in formation, which is the main multi-robot task addressed through our works, it has been investigated mainly two other cooperative tasks (namely: "Cooperative manipulation and transportation" and "Cooperative exploration under uncertainty"). These two tasks use mobile robotic entities and the necessity to coordinate their movements in order to optimize the mission achievement.

Afterward, the focus was on dynamic multi-robot navigation in formation and on the adopted cooperative strategies to perform safe, reliable and flexible navigation in cluttered environments. In the presented works, two main approaches have been emphasized:

- the "Virtual Structure" (VS) approach applied on Khepera® mobile robots (unicycle structure) and

- the "Leader-Follower" (LF) approach applied on VIPALAB® vehicles (tricycle structure).

For both approaches, a dedicated stable multi-controller architecture has been used based on target reaching/tracking. This strategy allows us to obtain a reactive architecture in the sense that the robots track the instantaneous state (pose and velocity) of their allocated virtual targets (thus, without any use of a reference trajectory or a trajectory planning process). An important part of this chapter focused on the way to ensure the formation reliability and stability. This was ensured while constraining the robots' set-points to become always attainable by the MRS. These formalizations permit us to obtain the maximal authorized dynamic of the formation while taking into account the robots' structural constraints and the targeted formation shape. The reliability of the achieved multi-robot task was evaluated according to a global Lyapunov function or Procrustes distances.

It was also presented in this chapter the adopted cooperative strategies to answer to the following questions:

q1 How do the robots determine their appropriate position in the formation?

q2 How do the robots act if there are obstructing obstacles (static/dynamic/others robots)?

q3 How does the MRS adapt dynamically its formation in order to deal with its environment (dynamic / configuration)?

To answer to these questions, the following was proposed:

a1 A dynamic targets allocation process called "Relative Cost Coefficient" (RCC). It corresponds to a cooperative and altruistic protocol between the robots in order to rapidly reach the formation. It permits us to have a decentralized and simple way to share the targets between the robots.

a2 Several techniques have been proposed to enhance the robots' safety:

- In the case of any identified dynamic obstacle, the obstacle avoidance strategy based on limit-cycles presented in chapter 2 has been enhanced. In its new version, it takes into account the obstacles' velocities to choose the most appropriate direction of avoidance.

- In the case of obstacle avoidance between robots of the same MRS, two mechanisms have been presented: the first is to impose one reference direction for all the MRS. Hence, when a robot detects a disturbing robot of the same group, it will always avoid it in a counter-clockwise direction (this can be seen as a local roundabout between them). The second enhances the inter-robot safety while defining a "penalty function" acting on the robots' linear velocities.

- In the case of obstacle avoidance, when the group is already in formation, it has been shown that the group can avoid the obstacle using three possible behaviors: each robot can locally avoid the obstacle using limit-cycles; or the MRS maintains its initial formation while using an appropriate limit-cycle with enough safe distance; the MRS switches to platoon formation the necessary time to safely avoid the obstacle.

a3 An adaptive and safe Strategy for Formation Reconfiguration (SFR) based on suitable smooth switches between different virtual target configurations. This strategy avoids the use of predefined trajectories and it can be applied in several situations when the formation has to be modified according to the environment context (dynamic, cluttered, etc.).

Several simulations and experiments, using either Khepera or VIPALAB robotic structures, highlighted the different proposals presented in this chapter.

General conclusion and prospects

GENERAL CONCLUSION

The presented research investigations are focused on the way to increase the autonomy of mobile a **mono robot** as well as **Multi-Robot Systems** (MRS) to achieve complex tasks. More precisely, the main objective is to propose appropriate **control architectures** in order to enhance the **safety, flexibility** and the **reliability** of autonomous navigations in complex environments (e.g., cluttered, uncertain and/or dynamic).

Although the developed concepts/methods/architectures could be applied for different domains (such as service robotics or agriculture), the **transportation domain** remains the privileged target. Applications include the transportation of persons (private car or public transport) as well as merchandise transportation (in a warehouse or port for instance).

The proposed control architectures (**decision/action**) have been addressed through three closely related elements: **task modeling** (according for instance to appropriate local or global reference frames); **planning** (short and long-term set-points generation) and finally **automatic control** aspects (stability and reliability for reaching the set-points). The above elements were gathered on appropriate **multi-controller architectures**, using a **bottom-up approach**, to tend ineluctably toward fully autonomous robot navigation even in highly dynamic and cluttered environments. In the overall presented works, the main objective is to develop autonomous mobile robots while using a **generic**, reliable and flexible process for modeling, planning and control. The proposed control architectures aim to deal online and safely with unpredictable/uncertain situations, and to optimize the overall robot navigation if the environment is well-known/mastered. It was highlighted also in our approach the importance to link the automatic control tools to those related to **Artificial Intelligence** (such as Markov Decision Process (**MDP**), Multi-Agent System (**MAS**) or **Fuzzy Logic**). In fact, even if the main targeted tasks concern autonomous navigation of vehicles for public transportation, which need a very high level of reliability, there exists always certain "navigation functions" (dealing for instance with the reasoning under uncertainties or to robots' coordination) which can be delegated to a higher level of abstraction and reasoning without losing the overall reliability of the control.

A large number of simulations and experiments using either Khepera® robots (unicycle structure) or VIPALAB® vehicles (tricycle structure) demonstrates the efficiency of the different proposals. More precisely, our contributions can be classified in 4 categories: (1) methodological certification; (2) obstacle avoidance based on PELC; (3) optimal short or long-term Trajectories/Waypoints planning; and finally (4) cooperative mobile multi-robot systems.

(1) Methodological certification

(1.1) Homogeneous and reliable multi-controller architectures

The use of multi-controller architectures is certainly the first characteristic of the adopted control methodology. Indeed, using this kind of control permits us to break up the complexity of the overall tasks to be carried out (by mono as well as by cooperative multi-robot systems) and allows therefore a bottom-up development (cf. section 1.4, page 15).

To demonstrate and certify the safety and reliability of the proposed multi-controller architectures, we had to develop/define several techniques/conventions/ideas:

- **Homogeneous set-points definition**: Based on target reaching/tracking. These set-points $T_S = (x_T, y_T, \theta_T, v_T)$ permit a generic and flexible way to define almost all mobile **robotic sub-tasks** (cf. section 2.4, page 37).

- **Stable control laws**: Once the format of set-points is defined, it is enough to develop appropriate stable control laws, which take into account the robot's structural constraints to stabilize the error asymptotically to zero. It is important to notice that the stability demonstrations have been proved while using mainly **Lyapunov synthesis**, and this for proposed elementary controllers (cf. section 3.2, page 54) as well as for the overall multi-controller architectures (cf. section 3.3, page 60).

- **Reliable elementary controllers**: To perform reliable robot navigation it is important to appropriately define the **elementary controllers** (behaviors) composing the multi-controller architecture (e.g., Leader tracking/following or obstacle avoidance). This is obtained while having a good balance among the following:

  - The set-points definition, which depends on the environment contexts (e.g., cluttered or not, dynamic or not, etc.). See for instance the influence of the R_S parameter (cf. section 2.4.2, page 38) which impacts the effective values of the obtained set-points T_S. R_S has been chosen $= 0$ in order to perform fully reactive obstacle avoidance (cf. motivation given in section 2.5.4.1 (page 43)).

  - The controllers law's parameters. See for instance section 5.3.1 (page 110) to determine the relation between the upper bound of the control errors and the controller parameters **K**.

- **Reliable controller coordination**: It is not enough to prove the stability of each elementary controller to guarantee the overall multi-controller architecture stability (cf. section 1.4.2, page 19). In fact, it is important also for such architectures to master the coordination between controllers' actions (hard switch or merging) in order to achieve smooth and reliable robot navigation. Several techniques were presented in section 3.3 (page 60) to master these controller interactions. Nevertheless, the most conclusive are those based on hard switching where the potentialities of **Hybrid**$_{CD}$ (Continuous/Discrete) systems have been taken as a formal framework to demonstrate analytically the overall architecture stability (while minimizing the controllers' set-points jerking). In this kind of controller coordination, two mechanisms of control, based on **Adaptive Functions** (cf. section 3.3.1) or **Adaptive Gains** (cf. section 3.3.2), have been presented and implemented on effective architectures.

It is interesting to notice that the homogeneous set-points definition permits us to have a common control law shared by several controllers. The proposed architectures enable us therefore to considerably simplify the stability analysis of the overall multi-controller architectures.

(1.2) Navigation through sequential waypoints

A new strategy was proposed based on successive static target reaching to perform reliable robot navigation even in a cluttered environment (cf. section 5.2, page 106). This strategy is an alternative (or a complementary) strategy to the widely used navigation approaches based on predefined reference trajectories. The main motivation of the proposed methodology arises from the need of further improving the navigation flexibility (to deal with different environments, tasks and contexts) while maintaining a high level of reliability and safety. Indeed, the use of only a discrete set of waypoints to guide the robot enables us to perform more maneuvers between waypoints (without the necessity of replanning any reference trajectory[1]), while remaining reliable, smooth (robot's set-points and trajectories) and safe (non-collision w.r.t. the road limits or any obstacle).

(1.3) Appropriate reference frames for task modeling/achievement

For simple and efficient description of robot navigation in any kind of environment, we proposed **tasks modeling** based on specific reference frames assigned for each obstacle / wall / target / etc. inside the considered environment (or at least for each element inside the robot's field of view). These **specific reference frames guide the robot behaviors** and permit us to **evaluate the success of the current achieved sub-task** (e.g., wall following, obstacle avoidance, target reaching/tracking, etc.). Each elementary reference frame therefore orientates locally the robot navigation. A kind of analogy could be established with robot manipulators modeling (cf. section 2.3.2,

---

[1]Which could be time-consuming and/or complex, mainly in cluttered and dynamic environments (cf. section 5.1).

page 35). The context of robot navigation is obviously different but the proposed reference frames help to make a reasoning on the efficiency of the robot movements in order to reach its final objective.

More precisely these reference frames permit in section 2.3.2 (page 35) to perform local and reactive obstacle avoidance-based PELC (Parallel Elliptic Limit-Cycle) and in section 4.3.2 (page 83) to optimize an overall global trajectory based on global PELC (gPELC). As emphasized also in the proposed strategy of navigation through sequential waypoints (cf. section 5.2, page 106) appropriate reference frames have been used to know when switching from one target to another.

(2) Obstacle avoidance based on PELC

The obstacle avoidance controller is one of the most important components to perform reliable robot navigation in cluttered and dynamic environments. It is the reason which led us to pay much more attention to the development of such controller. It has been proposed a reliable and flexible obstacle avoidance controller based on **generic orbital trajectories**, called PELC (**Parallel Elliptic Limit-Cycle**) and appropriate use of reference frames attributed to each static or dynamic obstacle in the environment (cf. section 2.3.2, page 35). The PELC enables us to bypass obstacles while keeping them always at a minimal distance (offset) (cf. section 2.3.1, page 31). It does not need any complex computation, from the moment that the parameters of the surrounded ellipse (characterizing the identified obstacles) are known. In addition, for the need of on-line obstacle avoidance controller several techniques to **detect** and to **characterize efficiently obstacles** were proposed (cf. section 2.5.2, page 40).

The use of the presented obstacle avoidance techniques based on limit-cycles has been proved to be very reliable and flexible to deal with different kinds of environments (e.g., cluttered or not, structured or not, static or dynamic). Several simulations and experiments showed its potentialities, for instance to perform reactive robot navigation (cf. section 2.5.5, page 47); hybrid (reactive/cognitive) navigation (cf. section 4.4.4, page 92) or for multi-vehicle navigation in cluttered environments (cf. sections 5.5.2 and 6.3.4.4, pages 135 and 170 respectively).

(3) Optimal short or long-term trajectories/waypoints planning

Several techniques of trajectories/waypoints planning were proposed (for both: short-term and long-term planning). Some of them use classical techniques (such as **Clothoids** or **Artificial Potential Fields**) which were adapted according to the new constraints/requirements of the achieved tasks (cf. section 4.2, page 79). The other part proposed techniques are relatively new in the sense that they use innovative components/methodology to plan the robot's movements. These techniques are based on **Multi-criteria optimization** and categorized as:

- **Waypoints generation**: To perform navigation based on sequential waypoints reaching (cf. section 5.2, page 106) optimal techniques have been proposed to optimize the generation of the set of waypoints (number, poses, velocities, etc.). Specifically an Optimal Multi-criteria Waypoint Selection based on Expanding Tree (**OMWS-ET**) or based on Grid-Map (**OMWS-GM**) were proposed (cf. section 5.4, (page 118)).

- **Optimal path planning-based PELC**: PELC can be used as an instantaneous planner to perform reactive obstacle avoidance (as given in section 2.5.4.1 (page 43)), but can be used also for more sophisticated avoidance / navigation. For short-term planning it was proposed the optimal PELC (**PELC***, (cf. section 4.3.1, page 80)) which is obtained while including several sub-criteria and constraints, among them the robot's initial state and structural constraints (non-holonomy and maximum steering); the enhancement of smoothness, safety of the obtained trajectories as well as the minimization of the robot's traveled distance. Furthermore, to perform appropriate cognitive navigation (when the overall environment is relatively well known), it is important to have a long-term planning technique. Hence, it was proposed to appropriately sequence a multitude of PELC* to obtain optimal global path planning-based <u>PELC*</u> (**gPELC***) (cf. section 4.3.2, page 83).

In addition, according specially to the possibility of using PELC as an instantaneous/short or long-term planner, it served us, as a main component to propose a **Hybrid**$_{RC}$ (Reactive/Cognitive) control architecture. More specifically, a <u>H</u>ybrid$_{RC}$ and <u>H</u>omogeneous (in term of set-points and control law) <u>C</u>ontrol <u>A</u>rchitecture (**HHCA** (cf. section 4.4, page 88)) was proposed. The main objective of this architecture is to permit us to deal on-line and safely with unpredictable/uncertain situations (reactive navigation), and to optimize the overall vehicle navigation if the environment is well-known/mastered (cognitive navigation). The proposed HHCA, with its different blocks and mechanisms, proved to be generic, flexible and reliable to deal with a large variety of environments (cluttered or not, structured or not and dynamic or not) while guaranteeing the **smoothness** of the switch between the robot's different navigation modes.

(4) Cooperative mobile multi-robot systems

Given notably the adopted multi-controller architecture (with its bottom-up construction) and the capitalization of knowledge, the already proposed techniques/components were naturally extended to deal with an even more complex system such as dynamic Cooperative Multi-Robot Systems (CMRS). Besides navigation in formation, which is the main multi-robot task addressed through our works, two other cooperative tasks were investigated (namely: "**Cooperative manipulation and transportation**" and "**Cooperative exploration under uncertainty**"). These two tasks share the fact of using <u>mobile</u> robotic entities and the necessity to coordinate their movements in order to optimize the mission achievement (cf. section 6.2, page 142).

To perform multi-robot navigation in formation, we used mainly two approaches: "**Virtual Structure**" (VS) (cf. section 6.3.3, page 149) and "**Leader-Follower**" (LF) (cf. section 6.3.4, page 160) to obtain safe, reliable and flexible cooperative navigation in cluttered environments. An important part of the proposed approaches made the focus on the way to ensure the formation reliability and stability. This was ensured notably while **constraining the robots' set-points** to become always attainable by the MRS. These formalizations therefore permit us to obtain the maximal authorized dynamic of the formation while taking into account the robots' structural constraints and the targeted formation shape.

In terms of cooperative strategies between the robots, several techniques / mechanisms were proposed such as:

- A decentralized approach to determine the robot's appropriate position in the formation which uses a specific dynamic targets allocation process called "Relative Cost Coefficient" (**RCC**).

- Several robots' safety functions proposed to better master the robot-robot coordination and the robot-"dynamic obstacles" avoidance (cf. section 6.3.3.3.2, page 155), such as extension of the technique of obstacle avoidance based on limit-cycles to **take into account the obstacles' velocities** or the "**penalty function**" to enhance inter-robot safety while acting on their linear velocities.

- An adaptive and safe Strategy for Formation Reconfiguration (**SFR**) based on suitable smooth switches between different virtual target configurations. This strategy avoids the use of predefined trajectories and it can be applied for different situations when the formation shape has to be modified dynamically according to the environment context (dynamic or not, cluttered or not, etc.).

PROSPECTS

The most important prospects are given summarily in what follows. For the long-term horizon the objective is to more fully standardize/homogenize the task of modeling and improve decisional/control processes in order to tend ineluctably toward fully autonomous robots (cf. section 1.2, page 7). Nowadays and according to the very rapid developments of software and hardware components, artificial life (as emphasized by Level 4 (cf. Figure 1.9, page 17)) could become a reality in a few decades.

More specifically it is targeted in the short and mid-term horizon to extend the potentialities of multi-controller architectures in order to gradually enhance the autonomy of mobile robots. The main elements which must be improved for multi-controller architectures correspond are:

- Demonstration of the overall control stability and reliability **even in the presence of uncertainties** (due for instance to the perceptions or to the robot modeling). The proposed mechanisms dealing with $Hybrid_{CD}$ systems will be extended while having always a formal framework to deal with such architecture. The degrees of robustness of the system to the noise/uncertainties should be quantified via suitable metrics in order to have a rigorous analysis of the control performances.

- Proposition of an analytic formulation of the optimal balance between reactive and cognitive robot navigation. This would be resolved notably by using appropriate metrics to better characterize the environment (dynamicity / uncertainty / etc.).

Simulation and experimental platforms

CONTENTS

A.1   KHEPERA ROBOT AND DEDICATED EXPERIMENTAL PLATFORM

For the need of our research for the cooperation of a group of mobile robots, an experimental platform based on 10 Khepera III mini-robots has been developed. The construction of this platform was motivated by the need to accelerate the experimental validation phases (rapid prototyping) monitoring / control architectures that will be investigated throughout our research. Some details of the experimental platform are given below.

Khepera III mobile robot   The Khepera® III mobile robot (cf. Figure A.1(a)) is a small unicycle mobile robot developed and distributed by K-Team[1] SA, Switzerland. This robot can handle additional modules such as cameras and grippers and can communicate using Wi-Fi or Bluetooth. It also has infrared proximity and ambient light sensors for environment interaction and obstacle avoidance.

Experimental platform   It has a dimension of $220cm \times 180cm$ (cf. Figure A.1(b)). The need to have an accurate tool to monitor the progress of the experiment led us to use a camera placed on top of the platform (cf. Figure A.1(c)). This allows a view of the entire experimental platform. Furthermore, to localize precisely (position and orientation) all the robots/obstacles in the environment, each entity has a

---

[1]http://www.k-team.com/mobile-robotics-products/old-products/
khepera-iii, consulted January 2015.

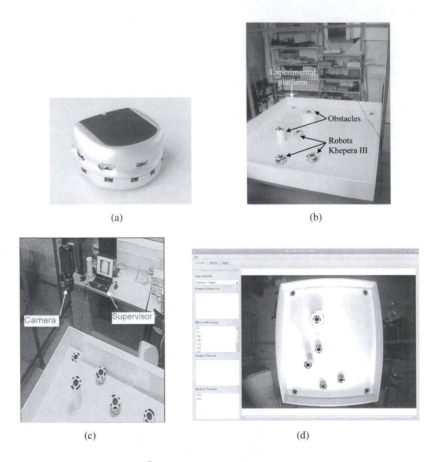

FIGURE A.1 (a) Khepera® III mobile robot (b) experimental platform, (c) used camera and supervisor unit, (d) supervisor software interface.

unique bar code [Lébraly et al., 2010] which permits us thanks to the top camera and to OpenCV® library, to accurately localize the robots and to transmit to them (using Wi-Fi) if necessary their pose and possibly (according to the experiment) the pose of other entities. More details about this experimental platform can be seen in [Benzerrouk, 2010, chapter 4]. Figure A.1(d) illustrates the used interface and the view of the environment from the central camera.

A.2 PIONEER ROBOT

Five Pioneer[®2] (cf. Figure A.2) robots have been used notably in the context of the R-Discover[3] project.

FIGURE A.2 Pioneer[®] mobile robot.

In addition to their ultrasonic sensors and odometers, these robots have been equipped with a fisheye camera and a laser sensor in order to better localize the robot in its environment [Lozenguez, 2012].

A.3 VIPALAB AND PAVIN PLATFORMS

The VIPALAB[®] from Apojee company [IP.Data.Sets, 2015] is a platform dedicated to the development of autonomous vehicles. This urban electric vehicle is used to implement several proposed control architectures for navigation in formation [Vilca et al., 2015a]. Some specifications of VIPALAB are shown in Table A.1

---

[2]http://www.mobilerobots.com/ResearchRobots/P3AT.aspx, consulted January 2015.

[3]http://home.mis.u-picardie.fr/~r-discover/doku.php?id=accueil, consulted January 2015.

TABLE A.1   VIPALAB platform.

VIPALAB	Elements	Description
	Chassis	(l, w, h)= $(1.96, 1.30, 2.11)\ m$
	Weight	$400\ kg$ (without batteries)
	Motor	Triphase $3x28\ V$, $4\ KW$
	Break	Integrated to the motor
	Maximum speed	$20\ km/h\ (\approx 5.5\ m/s)$
	Batteries	8 batteries $12\ V$, $80\ Ah$
	Autonomy	3 hours at full charge
	Computer	Intel Core i7, CPU:1.73 GHz RAM:8Go OS(32bits): Ubuntu12.04

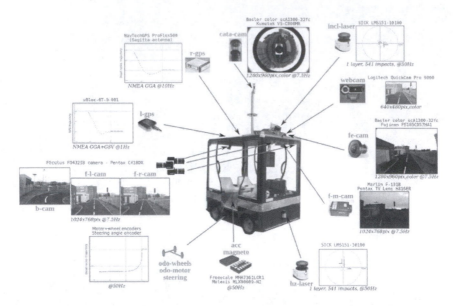

FIGURE A.3 VIPALAB with all sensors with their mounting locations and characteristics [IP.Data.Sets, 2015].

[IP.Data.Sets, 2015]. This vehicle carries different embedded proprioceptive and exteroceptive sensors such as cameras, odometers, gyrometer, steering angle sensor, an RTK-GPS, a Wi-Fi communication system and a computer (more details are given in [IP.Data.Sets, 2015]). The VIPALAB can be controlled using the on-board computer (through CAN protocol) or while using the wired control panel attached to the vehicle.

Several sensors are mounted in the VIPALAB to obtain information about the UGV or environment [IP.Data.Sets, 2015] (cf. Figure A.3). The main sensors used for experiments of the proposed control architecture for navigation in formation are described in Table A.2.

TABLE A.2 VIPALAB's sensors (cf. Figure A.3).

Elements	Description
RTK-GPS	NacTechGPS, accuracy: $2\ cm$ framerate: $10\ Hz$
IMU	Xsens MTi, accuracy: $0.2°/s$ framerate: $2\ KHz$
Range sensor	SICK LMS, range $[0,\ 50]\ m$ and angle $[-45°,\ 225°]$, resolution: $0.5°$ framerate: $50\ Hz$
Proprioceptive sensor	Wheel odometry, accuracy: $2\ cm$ framerate: $50\ Hz$
	Steering angle, resolution: $0.02°$ framerate: $50\ Hz$
	Motor odometry, resolution: $0.1\ m/s$ framerate: $50\ Hz$

Structured environment   The test platform named PAVIN (Plateforme d'Auvergne pour Véhicules Intelligents) is located at Campus Cézeaux of Blaise Pascal University in Clermont-Ferrand (cf. Figure A.4). PAVIN is an artificial environment composed of two areas (an urban and a rural area) which have a total ground surface of $5.000\ m^2$ which serves as a test-bed for mobile robotic applications. The urban area has a trajectory of $317\ m$ containing a scaled street with several traffic junctions and roundabouts with traffic sign boards and lights wherever necessary. Moreover, building facades on both sides, vegetation and street furniture are set to create a whole scene. The rural area has a trajectory of $264\ m$ with unpaved roads, grass and mud on the roadsides. In addition, the whole area is covered by a wireless network and a DGPS base station [IP.Data.Sets, 2015].

Although PAVIN is a small-scale environment, it stands as an ideal platform for evaluating algorithms related to autonomous driving such as navigation, road detection, traffic signal detection, etc. A 2D and a textured 3D model of the PAVIN environment geo-referenced with high-precision GPS data are available in [IP.Data.Sets, 2015].

FIGURE A.4   PAVIN experimental platform (Clermont-Ferrand, France).

A.4   ROBOTOPIA: REAL-TIME MAS SIMULATOR

ROBOTOPIA (cf. Figure A.5) is a real-time simulator using a multi-agent system (MAS) coordination. It was designed to simulate various scenarios to coordinate a group of mobile robots in cluttered environments.

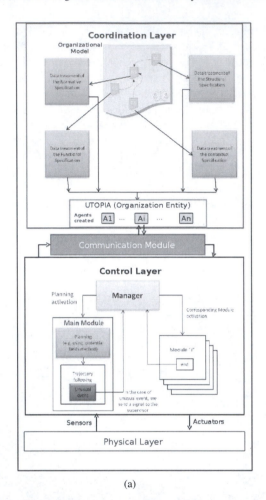

(a)

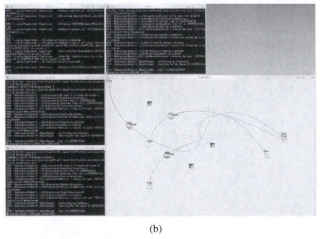

(b)

FIGURE A.5 (a) MAS Control architecture [Mouad et al., 2011b]. (b) Simulation of MRS using ROBOTOPIA [Mouad, 2014, chapter 6].

Stability of dynamic systems

CONTENTS

B.1   SYSTEM STABILITY

This appendix describes the definitions of stability applied to a dynamic system. The system is represented by the following state equation:

$$\dot{x} = f(x) \tag{B.1}$$

where $x \in \mathbb{R}^n$ represents the state of the system. The equilibrium point is assumed at $x = 0$.

- **Simple stability:** Let us consider that the initial time of the system is at $t_0 = 0$. Therefore, the origin point is stable if it satisfies the following expression:

$$\forall \epsilon > 0, \exists \delta > 0 : |x(0)| \leq \delta \Rightarrow |x(t)| \leq \epsilon \qquad \forall t \geq 0. \tag{B.2}$$

- **Asymptotic stability:** The system is asymptotically stable if it is stable and δ can be chosen such as:

$$|x(0)| \leq \delta \Rightarrow \lim_{t \to \infty} x(t) \to 0. \tag{B.3}$$

If eq. (B.3) is satisfied for all δ, then the system is globally asymptotically stable.

- **Exponential stability:** The system is exponentially stable if it satisfies:

$$\exists \delta > 0, c > 0, \lambda > 0 : |x(0)| \leq \delta \Rightarrow |x(t)| \leq c\,|x(0)|\,e^{-\lambda t} \ \forall t \geq 0. \tag{B.4}$$

B.2   STABILITY IN THE SENSE OF LYAPUNOV

First method: Indirect method

The first method of Lyapunov is based on the analysis of linearization of $f(x)$ system around its equilibrium point. This method consists of analyzing the eigenvalues $\lambda_i(J)$ of the Jacobian matrix J at its equilibrium point:

$$J = \frac{\partial f}{\partial x}(0). \tag{B.5}$$

The properties of the stability of systems are expressed as follows:

Theorem B.1 *First method of Lyapunov:*

1. *If all eigenvalues of the J matrix have a strictly negative real part, then the system is exponentially stable.*

2. *If the J matrix has at least one eigenvalue value with a strictly positive real part, then the system is unstable.*

If the system has at least one eigenvalue with a zero real part and any eigenvalue with a strictly positive real part, then no conclusion about the stability can be obtained. The system can be analyzed by the second method of Lyapunov.

Second method: Direct method

This method consists of a mathematical interpretation of an elementary observation: If the total energy of a system decreases/dissipates continuously over time, then the system tends to an equilibrium point, i.e., the system is stable. The idea is thus to find a temporal positive definite function which allows us to always have the negative derivative. This direct method is summarized in the next theorem.

Theorem B.2 *Second method of Lyapunov: The equilibrium point is stable if there exists a function V continuously differentiable and its derivative denoted by $\dot{V}$ satisfies:*

1. $V(0) = 0$,

2. $V(x) > 0 \quad \forall x \neq 0$,

3. $\dot{V}(x) \leq 0 \quad \forall x \neq 0$.

If condition (3) is replaced by $\dot{V}(x) < 0$, then the system is asymptotically stable.

Bibliography

[Abbadi et al., 2011] Abbadi, A., Matousek, R., and Petr Minar, P. S. (2011). RRTs review and options. In International Conference on Energy, Environment, Economics, Devices, Systems, Communications, Computers.

[ADAS, 2015] ADAS (2015). Advanced Driver Assistance Systems. http://en.wikipedia.org/wiki/Advanced_driver_assistance_systems, consulted January 2015.

[Adouane, 2005] Adouane, L. (2005). Architectures de controle comportementales et reactives pour la cooperation d'un groupe de robots mobiles. PhD thesis, Universite de Franche-Comte, LAB CNRS 6596.

[Adouane, 2008] Adouane, L. (2008). An adaptive multi-controller architecture for mobile robot navigation. In 10th IAS, Intelligent Autonomous Systems, 342–347, Baden-Baden, Germany.

[Adouane, 2009a] Adouane, L. (2009a). Hybrid and safe control architecture for mobile robot navigation. In 9th Conference on Autonomous Robot Systems and Competitions, Portugal.

[Adouane, 2009b] Adouane, L. (2009b). Orbital obstacle avoidance algorithm for reliable and on-line mobile robot navigation. Portuguese Journal Robotica N79, automacao. Selected from International Conference on Autonomous Robot Systems and Competitions.

[Adouane, 2010] Adouane, L. (2010). Cooperation bio-inspiree de systemes multi-robots autonomes. Editions universitaires europeennes edition.

[Adouane, 2013] Adouane, L. (2013). Towards smooth and stable reactive mobile robot navigation using on-line control set-points. In IEEE/RSJ, IROS'13, 5th Workshop on Planning, Perception and Navigation for Intelligent Vehicles, Tokyo-Japan.

[Adouane et al., 2011] Adouane, L., Benzerrouk, A., and Martinet, P. (2011). Mobile robot navigation in cluttered environment using reactive elliptic trajectories. In 18th IFAC World Congress, Milano-Italy.

[Adouane and Le Fort-Piat, 2004] Adouane, L. and Le Fort-Piat, N. (2004). Evolutionary parameters optimization for an hybrid control architecture of multicriteria tasks. In International Conference on Robotics and Biomimetics ROBIO, Shenyang-China. In CD, N°365.

[Adouane and Le-Fort-Piat, 2004] Adouane, L. and Le-Fort-Piat, N. (2004). Hybrid behavioral control architecture for the cooperation of minimalist mobile robots. In ICRA'04, International Conference on Robotics and Automation, 3735–3740, New Orleans-USA.

[Adouane and Le Fort-Piat, 2005] Adouane, L. and Le Fort-Piat, N. (2005). Methodology of parameters optimization for an hybrid architecture of control. In 16th IFAC World Congress, Prague-Czech Republic.

[Adouane and Le-Fort-Piat, 2006] Adouane, L. and Le-Fort-Piat, N. (2006). Behavioral and distributed control architecture of control for minimalist mobile robots. Journal Européen des Systèmes Automatisés, 40(2):177–196.

[Ahmadabadi and Nakano, 2001] Ahmadabadi, M. N. and Nakano, E. (2001). A "constrain and move" approach to distributed object manipulation. IEEE Transactions on Robotics and Automation, 17(2):157–172.

[Aicardi et al., 1995] Aicardi, M., Casalino, G., Bicchi, A., and Balestrino, A. (1995). Closed loop steering of unicycle like vehicles via Lyapunov techniques. Robotics Automation Magazine, IEEE, 2(1):27–35.

[Alami et al., 1998] Alami, R., Chatila, R., Fleury, S., Ghallab, M., and Ingrand, F. (1998). An architecture for autonomy. International Journal of Robotics Research, 17(4):315–337. Special Issue on Integrated Architectures for Robot Control and Programming.

[Albus, 1991] Albus, J. (1991). Outline for a theory of intelligence. IEEE Transactions on Systems, 21(3):473–509.

[Alfraheed and Al-Zaghameem, 2013] Alfraheed, M. and Al-Zaghameem, A. (2013). Exploration and cooperation robotics on the moon. Journal of Signal and Information Processing, 4:253–258.

[Anderson and Donath, 1990] Anderson, T. and Donath, M. (1990). Animal behavior as a paradigm for developing robot autonomy. Robotics and Autonomous Systems, 6:145–168.

[Antonelli et al., 2010] Antonelli, G., Arrichiello, F., and Chiaverini, S. (2010). The nsb control: A behavior-based approach for multi-robot systems. PALADYN Journal of Behavioral Robotics, 1:48–56.

[Arbib, 1981] Arbib, M. A. (1981). Perceptual structures and distributed motor control. In Handbook of Physiology, Section 2: The Nervous System, II, Motor Control, Part 1, 1449–1480.

[Arkin, 1989a] Arkin, R. (1989a). Towards the unification of navigational planning and reactive control. In AAAI Spring Symposium on Robot Navigation, 1–5.

[Arkin, 1989b] Arkin, R. C. (1989b). Motor schema-based mobile robot navigation. International Journal of Robotics Research, 8(4):92–112.

[Arkin, 1998] Arkin, R. C. (1998). Behavior-Based Robotics. The MIT Press.

[Arrúe et al., 1997] Arrúe, B., Cuesta, F., Braunstingl, R., and Ollero, A. (1997). Fuzzy behaviors combination to control a non-holonomic mobile robot using virtual perception memory. In Proceedings of the 6th IEEE International Conf. on Fuzzy Systems, 1239–1244, Barcelona, Spain.

[Aurenhammer, 1991] Aurenhammer, F. (1991). Voronoi diagrams – a survey of a fundamental geometric data structure. ACM Computing Surveys, 23(3):345–405.

[Bahr, 2009] Bahr, A. (2009). Cooperative Localization for Autonomous Underwater Vehicles. PhD thesis, Massachusetts Institute of Technology/Woods Hole Oceanographic Institution.

[Balch and Arkin, 1999] Balch, T. and Arkin, R. (1999). Behavior-based formation control for multi-robot teams. IEEE Transactions on Robotics and Automation.

[Baldassarre et al., 2003] Baldassarre, G., Nolfi, S., and Parisi, D. (2003). Evolution of collective behaviour in a team of physically linked robots. R. Gunther, A. Guillot, and J.-A. Meyer, Editors, Applications of Evolutionary Computing, Springer Verlag, Heidelberg, Germany, 581–592.

[Bellman, 1957] Bellman, R. (1957). A Markovian decision process. Journal of Mathematics and Mechanics, 6:679–684.

[Bellman et al., 1959] Bellman, R., Holland, J., and Kalaba, R. (1959). On an application of dynamic programming to the synthesis of logical systems. Journal of the ACM (JACM) archive, 6(4):486–493.

[Benzerrouk, 2010] Benzerrouk, A. (2010). Architecture de controle hybride pour systemes multi-robots : Application à la navigation en formation d'un groupe de robots mobiles. PhD thesis, Balise Pascal University.

[Benzerrouk et al., 2010a] Benzerrouk, A., Adouane, L., Al-Barakeh, Z., and Martinet, P. (2010a). Stabilite globale pour la navigation reactive d'un robot mobile en presence d'obstacles. In CIFA 2010, Sixieme Conference Internationale Francophone d'Automatique, Nancy-France.

[Benzerrouk et al., 2010b] Benzerrouk, A., Adouane, L., Lequievre, l., and Martinet, P. (2010b). Navigation of multi-robot formation in unstructured environment using dynamical virtual structures. In IROS'10, IEEE/RSJ International Conference on Intelligent Robots and Systems, 5589–5594, Taipei-Taiwan.

[Benzerrouk et al., 2010c] Benzerrouk, A., Adouane, L., and Martinet, P. (2010c). Lyapunov global stability for a reactive mobile robot navigation in presence of obstacles. In ICRA'10 International Workshop on Robotics and Intelligent Transportation System, RITS10, Anchorage-Alaska.

[Benzerrouk et al., 2012a] Benzerrouk, A., Adouane, L., and Martinet, P. (2012a). Altruistic distributed target allocation for stable navigation in formation of multi-robot system. In 10th International IFAC Symposium on Robot Control (SYROCO'12), Dubrovnik - Croatia.

[Benzerrouk et al., 2012b] Benzerrouk, A., Adouane, L., and Martinet, P. (2012b). Dynamic obstacle avoidance strategies using limit cycle for the navigation of multi-robot system. In 2012 IEEE/RSJ IROS'12, 4th Workshop on Planning, Perception and Navigation for Intelligent Vehicles, Vilamoura, Algarve, Portugal.

[Benzerrouk et al., 2013] Benzerrouk, A., Adouane, L., and Martinet, P. (2013). Obstacle avoidance controller generating attainable set-points for the navigation of multi-robot system. In IEEE Intelligent Vehicles Symposium (IV), Gold Coast - Australia.

[Benzerrouk et al., 2014] Benzerrouk, A., Adouane, L., and Martinet, P. (2014). Stable navigation in formation for a multi-robot system based on a constrained virtual structure. Robotics and Autonomous Systems (RAS), 62(12):1806–1815.

[Benzerrouk et al., 2008] Benzerrouk, A., Adouane, L., Martinet, P., and Andreff, N. (2008). Toward an hybrid control architecture for a mobile multi-robot systems. In 3rd National Conference on Control Architectures of Robots (CAR08), Bourges - France.

[Benzerrouk et al., 2009] Benzerrouk, A., Adouane, L., Martinet, P., and Andreff, N. (2009). Multi Lyapunov function theorem applied to a mobile robot tracking a trajectory in presence of obstacles. In European Conference on Mobile Robots (ECMR 2009), Milini/Dubrovnik Croatia.

[Berhault et al., 2003] Berhault, M., Huang, H., Keskinocak, P., Koenig, S., Elmaghraby, W., Griffin, P. M., and Kleywegt, A. J. (2003). Robot exploration with combinatorial auctions. In IEEE/RSJ International Conference on Intelligent Robots and Systems, 1957–1962, Las Vegas, Nevada, USA.

[Bertsekas, 1995] Bertsekas, D. P. (1995). Dynamic Programming and Optimal Control, volume 1. Belmont Massachusetts, Athena Scientific.

[Blazic, 2012] Blazic, S. (2012). Four-state trajectory-tracking control law for wheeled mobile robots. In 10th International IFAC Symposium on Robot Control. Croatia.

[Bonabeau et al., 1999] Bonabeau, E., Dorigo, M., and Theraulaz, G. (1999). Swarm Intelligence: From Natural to Artificial Systems. Oxford University Press.

[Bondy and Murty, 2008] Bondy, J. and Murty, U. (2008). Graph theory. Graduate Texts in Mathematics 244. Berlin: Springer. xii, 651 p.

[Bonfè et al., 2012] Bonfè, M., Secchi, C., and Scioni, E. (2012). Online trajectory generation for mobile robots with kinodynamic constraints and embedded control systems. In 10th International IFAC Symposium on Robot Control. Croatia.

[Borja et al., 2013] Borja, R., de la Pinta, J., Álvarez, A., and Maestre, J. (2013). Integration of service robots in the smart home by means of UPnP: A surveillance robot case study. Robotics and Autonomous Systems, 61(2):153–160.

[Boufera et al., 2014] Boufera, F., Debbat, F., Adouane, L., and Khelfi Faycal, K. M. (2014). Mobile robot navigation using fuzzy limit-cycles in cluttered environment. International Journal of Intelligent Systems and Applications, (7):12–21.

[Braitenberg, 1984] Braitenberg, V. (1984). Vehicles: Experiments in Synthetic Psychology. Cambridge, MA: MIT Press.

[Branicky, 1993] Branicky, M. S. (1993). Stability of switched and hybrid systems. In 33rd IEEE Conference on Decision Control, 3498–3503, USA.

[Branicky, 1998] Branicky, M. S. (1998). Multiple Lyapunov functions and other analysis tools for switched and hybrid systems. IEEE Transaction on Automatic Control, 43(4):475–482.

[Brock and Khatib, 1999] Brock, O. and Khatib, O. (1999). High-speed navigation using the global dynamic window approach. In IEEE International Conference on Robotics and Automation, 341–346, Detroit, Michigan, USA.

[Brogliato et al., 1997] Brogliato, B., Niculescu, S., and Orhant (1997). On the control of finite dimensional mechanical systems with unilateral constraints. IEEE Transactions on Automatic Control, 42(2):200–215.

[Brooks, 1986] Brooks, R. A. (1986). A robust layered control system for a mobile robot. IEEE Journal of Robotics and Automation, RA-2:14–23.

[Buehler et al., 2009] Buehler, M., Iagnemma, K., and Singh, S. (2009). The DARPA urban challenge: Autonomous vehicles in city traffic, volume 56. Springer Science & Business Media.

[Burns, 2013] Burns, L. D. (2013). Sustainable mobility: A vision of our transport future. Nature, 497(7448):181–182.

[Busquets et al., 2003] Busquets, D., Sierra, C., and de Mántaras, R. L. (2003). A multiagent approach to qualitative landmark-based navigation. Autonomous Robots, 15(2):129–154.

[Cao et al., 1997] Cao, Y., Fukunaga, A. S., and Kahng, A. B. (1997). Cooperative mobile robotics: Antecedents and directions. Autonomous Robots, 4:1–23.

[Caprari, 2003] Caprari, G. (2003). Autonomous micro-robots: Applications and limitations. PhD thesis, Faculté Sciences et Techniques de l'Ingénieur, École Polytechnique Fédérale de Lausanne. Thèse N°2753.

[Cariou et al., 2010] Cariou, C., Lenain, R., Thuilot, B., and Martinet, P. (2010). Autonomous maneuver of a farm vehicle with a trailed implement: Motion planner and lateral-longitudinal controllers. In IEEE International Conference on Robotics and Automation, 3819–3824, Anchorage, Alaska, USA.

[Causse and Pampagnin, 1995] Causse, O. and Pampagnin, L. (1995). Management of a multi-robot system in a public environment. In Proceedings of IEEE/RSJ International Conference on Intelligent Robots and Systems, IROS, 246–252, Pittsburgh, PA, USA.

[Chao et al., 2012] Chao, Z., Zhou, S.-L., Ming, L., and Zhang, W.-G. (2012). UAV formation flight based on nonlinear model predictive control. Mathematical Problems in Engineering, 2012:1–15.

[Chen and Li, 2006] Chen, X. and Li, Y. (2006). Smooth formation navigation of multiple mobile robots for avoiding moving obstacles. International Journal of Control, Automation, 4(4):466–479.

[Choset et al., 2005] Choset, H., Lynch, K., Hutchinson, S., Kantor, G., Burgard, W., Kavraki, L., and Thrun, S. (2005). Principles of Robot Motion: Theory, Algorithms, and Implementation. MIT Press.

[Connell, 1990] Connell, J. H. (1990). Minimalist Mobile Robotics. Academic Press, Londres.

[Connors and Elkaim, 2007] Connors, J. and Elkaim, G. H. (2007). Manipulating b-spline based paths for obstacle avoidance in autonomous ground vehicles. In ION National Technical Meeting, ION NTM 2007, San Diego, CA, USA.

[Consolini et al., 2008] Consolini, L., Morbidi, F., Prattichizzo, D., and Tosques, M. (2008). Leader-follower formation control of nonholonomic mobile robots with input constraints. Automatica, 44(5):1343–1349.

[Dafflon et al., 2015] Dafflon, B., Vilca, J., Gechter, F., and Adouane, L. (2015). Adaptive autonomous navigation using reactive multi-agent system for control laws merging. Procedia Computer Science. Selected paper from the International Conference on Computational Science ICCS'15 (Reykjavík, Iceland).

[DARPA, 2015] DARPA (2015). Grand Challenge. https://fr.wikipedia.org/wiki/DARPA_Grand_Challenge, consulted January 2015.

[Das et al., 2002] Das, A., Fierro, R., Kumar, V., Ostrowski, J., Spletzer, J., and Taylor, C. (2002). A vision-based formation control framework. IEEE Transaction on Robotics and Automation, 18(5):813–825.

[Dasgupta et al., 2011] Dasgupta, P., Whipple, T., and Cheng, K. (2011). Effects of multi-robot team formations on distributed area coverage. International Journal of Swarm Intelligent Research, 2(1):44–69.

[Daviet and Parent, 1997] Daviet, P. and Parent, M. (1997). Platooning for small public urban vehicles. In Khatib, O. and Salisbury, J., editors, Experimental Robotics IV, volume 223 of Lecture Notes in Control and Information Sciences, 343–354. Springer Berlin Heidelberg.

[De Maesschalck et al., 2000] De Maesschalck, R., Jouan-Rimbaud, D., and Massart, D. (2000). The Mahalanobis distance. Chemometrics and Intelligent Laboratory Systems, 50(1):1–18.

[Denavit and Hartenberg, 1955] Denavit, J. and Hartenberg, R. S. (1955). A kinematic notation for lower-pair mechanisms based on matrices. Trans. ASME E, Journal of Applied Mechanics, 22:215–221.

[Desai et al., 2001] Desai, J., Ostrowski, J., and Kumar, V. (2001). Modeling and control of formations of nonholonomic mobile robots. IEEE Transaction on Robotics and Automation, 17(6):905–908.

[Dijkstra, 1959] Dijkstra, E. W. (1959). A note on two problems in connexion with graphs. Numerische Mathematik, 1:269–271.

[Do, 2007] Do, K. D. (2007). Formation tracking control of unicycle-type mobile robots. In IEEE International Conference on Robotics and Automation, 2391–2396, Roma, Italy.

[Drogoul, 1993] Drogoul, A. (1993). De la simulation multi-agent à la résolution collective de problèmes : Une étude de l'émergence de structures d'organisation dans les systèmes multi-agents. PhD thesis, Université Paris VI.

[Egerstedt, 2000] Egerstedt, M. (2000). Behavior based robotics using hybrid automata. In Lynch, N. and Krogh, B., editors, Hybrid Systems: Computation and Control, volume 1790 of Lecture Notes in Computer Science, 103–116. Springer Berlin Heidelberg.

[Egerstedt and Hu, 2002] Egerstedt, M. and Hu, X. (2002). A hybrid control approach to action coordination for mobile robots. Automatica, 38(1):125–130.

[El Jalaoui et al., 2005] El Jalaoui, A., Andreu, D., and Jouvencel, B. (2005). A control architecture for contextual tasks management: Application to the AUV taipan. In Oceans 2005 - Europe, volume 2, 752–757.

[El-Zaher et al., 2012] El-Zaher, M., Contet, J.-M., Gechter, F., and Koukam, A. (2012). Echelon platoon organisation: A distributed approach based on 2-spring virtual links. In Proceeding of the 15th International Conference on Artificial Intelligence: Methodology, Systems, Applications, Germany.

[Eskandarian, 2012] Eskandarian, A. (2012). Handbook of Intelligent Vehicles. Springer edition.

[EUREKA, 1995] EUREKA, E. (1995). http://www.eurekanetwork.org/project/-/id/45.

[Fagnant and Kockelman, 2015] Fagnant, D. J. and Kockelman, K. (2015). Preparing a nation for autonomous vehicles: opportunities, barriers and policy recommendations. Transportation Research Part A: Policy and Practice, 77:167–181.

[Fang et al., 2005] Fang, H., Lenain, R., Thuilot, B., and Martinet, P. (2005). Trajectory tracking control of farm vehicles in presence of sliding. In IEEE/RSJ International Conference on Intelligent Robots and Systems, 58–63, Edmonton, Alberta, Canada.

[Fernández-Madrigal and Claraco, 2013] Fernández-Madrigal, J.-A. and Claraco, J. L. B. (2013). Simultaneous Localization and Mapping for Mobile Robots: Introduction and Methods. IGI Global, Hershey, PA, USA.

[Ferrell, 1995] Ferrell, C. (1995). Global behavior via cooperative local control. Autonomous Robots, 2(2):105–125.

[Firby, 1987] Firby, R. (1987). An investigation into reactive planning in complex domains. In 6th National Conference on Artificial Intelligence, 202–206, Seattle.

[Fleury et al., 1993] Fleury, S., Souères, P., Laumond, J.-P., and Chatila, R. (1993). Primitives for smoothing mobile robot trajectories. In ICRA, 832–839.

[Fraichard, 1999] Fraichard, T. (1999). Trajectory planning in a dynamic workspace: A "state time" approach. Advanced Robotics, 13(1):75–94.

[Franco et al., 2015] Franco, C., Stipanovic, D. M., Lopez-Nicolas, G., Sagues, C., and Llorente, S. (2015). Persistent coverage control for a team of agents with collision avoidance. European Journal of Control, 22:30–45.

[Fukuda et al., 2000] Fukuda, T., Takagawa, I., Sekiyama, K., and Hasegawa, Y. (2000). chapter Hybrid Approach of Centralized Control and Distributed Control for Flexible Transfer System, 65–85. Kluwer Academic Publishers.

[Gasser et al., 2013] Gasser, T. M., Arzt, C., Ayoubi, M., Bartels, A., Burkle, L., Eier, J., Flemisch, F., Hacker, D., Hesse, T., Huber, W., Lotz, C., Maurer, M., Ruth-Schumacher, S., Schwarz, J., and Vogt, W. (2013). Legal consequences of an increase in vehicle automation. (English translation) Bundesanstalt fur Stra. (BASt), BASt-Report F83 (Part 1).

[Gat, 1992] Gat, E. (1992). Integrating planning and reacting in a heterogeneous asynchronous architecture for controlling real-world mobile robots. In Proceedings of AAAI-92, 809–815, San Jose, CA, USA.

[Gat, 1998] Gat, E. (1998). Three-layer architecture. In D. Kortenkamp, R. B. and Murphy, R., editors, Artificial Intelligence and Mobile Robotics, 195–210. AAAI Press.

[GCDC, 2016] GCDC (2016). Grand Cooperative Driving Challenge. http://www.gcdc.net/, consulted January 2015.

[Gerkey and Mataric', 2002] Gerkey, P. and Mataric', M. J. (2002). Sold!: Auction methods for multirobot coordination. IEEE Transactions on Robotics and Automation, 18:758–768.

[Ghommam et al., 2010] Ghommam, J., Mehrjerdi, H., Saad, M., and Mnif, F. (2010). Formation path following control of unicycle-type mobile robots. Robotics and Autonomous Systems, 58(5):727–736.

[Gim et al., 2014a] Gim, S., Adouane, L., Lee, S., and Derutin, J.-P. (2014a). Parametric continuous curvature trajectory for smooth steering of the car-like vehicle. In 13th International Conference on Intelligent Autonomous System (IAS-13), Padova -Italy.

[Gim et al., 2014b] Gim, S., Adouane, L., Lee, S., and Derutin, J.-P. (2014b). Smooth trajectory generation with 4d space analysis for dynamic obstacle avoidance. In 11th International Conference on Informatics in Control, Automation and Robotics (ICINCO'14), Vienna - Austria. Special Session on Intelligent Vehicle Controls and Intelligent Transportation Systems - IVC-ITS 2014.

[Goerzen et al., 2010] Goerzen, C., Kong, Z., and Mettler, B. (2010). A survey of motion planning algorithms from the perspective of autonomous UAV guidance. J. Intell. Robotics Syst., 57(1-4):65–100.

[Goodrich and Schultz, 2007] Goodrich, M. A. and Schultz, A. C. (2007). Human-robot interaction: A survey. Found. Trends Hum.-Comput. Interact., 1(3):203–275.

[GoogleCar, 2015] GoogleCar (2015). http://en.wikipedia.org/wiki/Google_driverless_car, consulted January 2015.

[Grassi Junior et al., 2006] Grassi Junior, V., Parikh, S., and Okamoto Junior, J. (2006). Hybrid deliberative/reactive architecture for human-robot interaction. ABCM Symposium Series in Mechatronics, 2:563–570.

[Gu and Dolan, 2012] Gu, T. and Dolan, J. M. (2012). On-road motion planning for autonomous vehicles. In Su, C.-Y., Rakheja, S., and Liu, H., editors, Intelligent Robotics and Applications, volume 7508. Springer Berlin Heidelberg.

[Guillet et al., 2014] Guillet, A., Lenain, R., Thuilot, B., and Martinet, P. (2014). Adaptable robot formation control: Adaptive and predictive formation control of autonomous vehicles. IEEE Robotics & Automation Magazine, 21(1):28–39.

[Gulati, 2011] Gulati, S. (2011). A framework for characterization and planning of safe, comfortable, and customizable motion of assistive mobile robots. PhD thesis, The University of Texas at Austin.

[Gustavi and Hu, 2008] Gustavi, T. and Hu, X. (2008). Observer-based leader-following formation control using onboard sensor information. IEEE Transactions on Robotics, 24:1457–1462.

[Güttler et al., 2014] Güttler, J., Georgoulas, C., Linner, T., and Bock, T. (2014). Towards a future robotic home environment: A survey. Gerontology (International Journal of Experimental, Clinical, Behavioural and Technological Gerontology).

[Harary, 1969] Harary, F. (1969). Graph Theory. Addison-Wesley, Reading, MA.

[Hichri et al., 2014a] Hichri, B., Adouane, L., Fauroux, J.-C., Mezouar, Y., and Doroftei, I. (2014a). Cooperative lifting and transport by a group of mobile robots. In Springer Tracts in Advanced Robotics, from International Symposium on Distributed Autonomous Robotic Systems, DARS 2014, Daejeon - Korea.

[Hichri et al., 2014b] Hichri, B., Fauroux, J.-C., Adouane, L., Doroftei, I., and Mezouar, Y. (2014b). Lifting mechanism for payload transport by collaborative mobile robots. In 5th European Conference on Mechanism Science (EUCOMES), Guimaraes, Portugal.

[Hichri et al., 2015] Hichri, B., Fauroux, J.-C., Adouane, L., Doroftei, I., and Mezouar, Y. (2015). Lifting mechanism for payload transport by collaborative mobile robots. New Trends in Mechanism and Machine Science - Mechanisms and Machine Science, 24:157–165. Selected paper from International EUCOMES conference.

[Hichri et al., 2014c] Hichri, B., Fauroux, J.-C., Adouane, L., Mezouar, Y., and Doroftei, I. (2014c). Design of collaborative cross and carry mobile robots c3bots. Advanced Materials Research, 837:588 – 593. Ed. Springer.

[Hirata et al., 2002] Hirata, Y., Hatsukari, T., Kosuge, K., Asama, H., Kaetsu, H., and Kawabata, K. (2002). Transportation of an object by multiple distributed robot helpers in cooperation with a human. Transactions of the Japan Society of Mechanical Engineers, 68(668):181–188.

[Hirata et al., 2007] Hirata, Y., Matsuda, Y., and Kosuge, K. (2007). Handling of an object in 3-D space by multiple mobile manipulators based on intentional force/moment applied by human. In IEEE/ASME International Conference on Advanced Intelligent Mechatronics, 1–6.

[Horst and Barbera, 2006] Horst, J. and Barbera, A. (2006). Trajectory generation for an on-road autonomous vehicle.

[Hsu and Liu, 2007] Hsu, H. C.-H. and Liu, A. (2007). A flexible architecture for navigation control of a mobile robot. IEEE Transactions on Systems, Man, and Cybernetics, Part A, 37(3):310–318.

[Huntsberger et al., 2003] Huntsberger, T., Pirjanian, P., Trebi-ollennu, A., Nayar, H. A., Ganino, A. J., Garrett, M., Joshi, S., and Schenker, S. P. (2003). Campout: A control architecture for tightly coupled coordination of multi-robot systems for planetary surface exploration. IEEE Trans. Systems, Man & Cybernetics, Part A: Systems and Humans, 33:550–559.

[Ider, 2009] Ider, M. (2009). Controle/commande par logique floue de la coopera-tion d'un groupe de robots mobiles pour la navigation en convoi. Master's thesis, MASTER II STIC: Informatique MSIR-RPM-UBP.

[IP.Data.Sets, 2015] IP.Data.Sets, S. (2015). http://ipds.univ-bpclermont.fr.

[Jie et al., 2006] Jie, M. S., Baek, J. H., Hong, Y. S., and Lee, K. W. (2006). Real time obstacle avoidance for mobile robot using limit-cycle and vector field method. Knowledge-Based Intelligent Information and Engineering Systems.

[Johansson et al., 1999] Johansson, K. H., M., E., J., L., and S., S. (1999). On the regularization of zeno hybrid automata. Systems & Control Letters, 38:141–150.

[Jones and Snyder, 2001] Jones, H. L. and Snyder, M. (2001). Supervisory control of multiple robots based on real-time strategy game interaction paradigm. In International Conference on Systems, Man and Cybernetics, 383–388.

[Kallem et al., 2011] Kallem, V., Komoroski, A., and Kumar, V. (2011). Sequential composition for navigating a nonholonomic cart in the presence of obstacles. IEEE Transactions on Robotics, 27(6):1152–1159.

[Kanayama et al., 1990] Kanayama, Y., Kimura, Y., Miyazaki, F., and Noguchi, T. (1990). A stable tracking control method for an autonomous mobile robot. In Proceedings of the IEEE International Conference on Robotics and Automation, 384–389.

[Karaman and Frazzoli, 2011] Karaman, S. and Frazzoli, E. (2011). Sampling-based algorithms for optimal motion planning. International Journal of Robotics Research, 30(7):846–894.

[Kendall, 1989] Kendall, D. G. (1989). A survey of the statistical theory of shape. Statistical Science, 4:87–99.

[Kenyon and Morton, 2003] Kenyon, A. S. and Morton, D. P. (2003). Stochastic vehicle routing with random travel times. Transportation Science, 37(1):69–82.

[Khalil, 2002] Khalil, H. K. (2002). Nonlinear Systems. 3rd edition.

[Khalil and Dombre, 2004] Khalil, W. and Dombre, E. (2004). Modeling, Identification and Control of Robots. Hermes Penton.

[Khansari-Zadeh and Billard, 2012] Khansari-Zadeh, S.-M. and Billard, A. (2012). A dynamical system approach to real-time obstacle avoidance. Autonomous Robots, 32:433–454. 10.1007/s10514-012-9287-y.

[Khatib, 1986] Khatib, O. (1986). Real-time obstacle avoidance for manipulators and mobile robots. The International Journal of Robotics Research, 5:90–99.

[Kim and Kim, 2003] Kim, D.-H. and Kim, J.-H. (2003). A real-time limit-cycle navigation method for fast mobile robots and its application to robot soccer. Robotics and Autonomous Systems, 42(1):17–30.

[Klančar et al., 2011] Klančar, G., Matko, D., and Blažič, S. (2011). A control strategy for platoons of differential drive wheeled mobile robot. Robotics and Autonomous Systems, 59(2):57–64.

[Konolige et al., 1997] Konolige, K., Myers, K., Ruspini, E., and Saffiotti, A. (1997). The Saphira architecture: A design for autonomy. Journal of Experimental and Theoretical Artificial Intelligence, 9:215–235.

[Kruppa et al., 2000] Kruppa, H., Fox, D., Burgard, W., and Thrun., S. (2000). A probabilistic approach to collaborative multirobot localization. Autonomous Robots, 8:325–344.

[Kuwata et al., 2008] Kuwata, Y., Fiore, G. A., Teo, J., Frazzoli, E., and How, J. P. (2008). Motion planning for urban driving using RRT. In Proceedings of the IEEE International Conference on Intelligent Robots and Systems, 1681–1686, Nice, France.

[Labakhua et al., 2008] Labakhua, L., Nunes, U., Rodrigues, R., and Leite, F. (2008). Smooth trajectory planning for fully automated passengers vehicles: Spline and clothoid based methods and its simulation. In Cetto, J., Ferrier, J.-L., Costa dias Pereira, J. M., and Filipe, J., editors, Informatics in Control Automation and Robotics, volume 15 of Lecture Notes Electrical Engineering, 169–182. Springer Berlin Heidelberg.

[Lategahn et al., 2011] Lategahn, H., Geiger, A., and Kitt, B. (2011). Visual SLAM for autonomous ground vehicles. In IEEE International Conference on Robotics and Automation, 1732–1737. IEEE.

[Latombe, 1991] Latombe, J.-C. (1991). Robot Motion Planning. Kluwer Academic Publishers, Boston, MA.

[Laumond, 2001] Laumond, J.-P. (2001). La Robotique Mobile. Traité IC2 Information-Commande-Communication. Hermès.

[Lavalle, 1998] Lavalle, S. M. (1998). Rapidly-exploring random trees: A new tool for path planning. Technical report, Computer Science Dept., Iowa State University.

[LaValle, 2006] LaValle, S. M. (2006). Planning Algorithms. Cambridge Univ. Press.

[Lébraly et al., 2010] Lébraly, P., Deymier, C., Ait-Aider, O., Royer, E., and Dhome, M. (2010). Flexible extrinsic calibration of non-overlapping cameras using a planar mirror: Application to vision-based robotics. IEEE International Conference on Intelligent Robots and Systems, 5640–5647.

[Léchevin et al., 2006] Léchevin, N., Rabbath, C., and Sicard, P. (2006). Trajectory tracking of leader-follower formations characterized by constant line-of-sight angles. Automatica, 42:2131–2141.

[Lee and Litkouhi, 2012] Lee, J.-W. and Litkouhi, B. (2012). A unified framework of the automated lane centering/changing control for motion smoothness adaptation. In 15th International IEEE Conference on Intelligent Transportation Systems (ITSC), 282–287.

[Levinson and Thrun, 2010] Levinson, J. and Thrun, S. (2010). Robust vehicle localization in urban environments using probabilistic maps. In IEEE International Conference on Robotics and Automation. Alaska, USA.

[Li et al., 2005] Li, X., Xiao, J., and Cai, Z. (2005). Backstepping based multiple mobile robots formation control. In IEEE/RSJ International Conference on Intelligent Robots and Systems, 887–892, Edmonton, Alberta, Canada. IEEE.

[Liberzon, 2003] Liberzon, D. (2003). Switching in Systems and Control. Birkhauser.

[Litman, 2013] Litman, T. (2013). Autonomous Vehicle Implementation Predictions: Implications for Transport Planning.

[Lozenguez, 2012] Lozenguez, G. (2012). Stratégies coopératives pour l'exploration et la couverture spatiale pour une flotte de robots explorateurs. PhD thesis, Université de Caen.

[Lozenguez et al., 2011a] Lozenguez, G., Adouane, L., Beynier, A., Martinet, P., and Mouaddib, A. I. (2011a). Calcul distribue de politiques d'exploration pour une flotte de robots mobiles. In JFSMA'11, 19eme Journees Francophones sur les Systemes Multi-Agents, Valenciennes-France.

[Lozenguez et al., 2011b] Lozenguez, G., Adouane, L., Beynier, A., Martinet, P., and Mouaddib, A. I. (2011b). Map partitioning to approximate an exploration strategy in mobile robotics. In PAAMS 2011, 9th International Conference on Practical Applications of Agents and Multi-Agent Systems, Salamanca-Spain. Published after in Advances on Practical Applications of Agents and Multiagent Systems Advances in Intelligent and Soft Computing, 88, 2011,63–72.

[Lozenguez et al., 2013a] Lozenguez, G., Adouane, L., Beynier, A., Martinet, P., and Mouaddib, A. I. (2013a). Resolution approchee par decomposition de processus decisionnels de markov appliquee a l'exploration en robotique mobile. In JFPDA, 8emes Journees Francophones sur la Planification, la Decision et l'Apprentissage pour la conduite de systemes, Lille - France.

[Lozenguez et al., 2012a] Lozenguez, G., Adouane, L., Beynier, A., Mouaddib, A. I., and Martinet, P. (2012a). Interleaving planning and control of mobiles robots in urban environments using road-map. In 12th International Conference on Intelligent Autonomous System (IAS-12), Jeju Island - Korea. Published after in Advances in Intelligent Systems and Computing, 193, 2013, 683-691.

[Lozenguez et al., 2012b] Lozenguez, G., Adouane, L., Beynier, A., Mouaddib, A. I., and Martinet, P. (2012b). Map partitioning to approximate an exploration strategy in mobile robotics. Multiagent and Grid Systems (MAGS), 8(3):275–288.

[Lozenguez et al., 2013b] Lozenguez, G., Beynier, A., Adouane, L., Mouaddib, A. I., and Martinet, P. (2013b). Simultaneous auctions for "rendez-vous" coordination phases in multi-robot multi-task mission. In IAT13, IEEE/WIC/ACM International Conference on Intelligent Agent Technology, Atlanta, GA USA.

[Lu and Shladover, 2014] Lu, X.-Y. and Shladover, S. (2014). Automated truck platoon control and field test. In Meyer, G. and Beiker, S., editors, Road Vehicle Automation, Lecture Notes in Mobility, 247–261. Springer International Publishing.

[Luca et al., 1998] Luca, A. D., Oriolo, G., and Samson, C. (1998). Feedback control of a nonholonomic car-like robot. In Laumond, J.-P., editor, Robot Motion Planning and Control, 171–253. Springer-Verlag.

[Maes, 1989] Maes, P. (1989). The dynamics of action selection. In Proceedings of the Eleventh International Joint Conference on Artificial Intelligence (IJCAI), 991–97, Detroit.

[Maes, 1991] Maes, P. (1991). A bottom-up mechanism for action selection in an artificial creature. In by S. Wilson, E. and Arcady-Meyer, J., editors, From Animals to Animats: Proceedings of the Adaptive Behavior Conference, 238–246, Paris-France. MIT Press.

[Martins et al., 2012] Martins, M. M., Santos, C. P., Frizera-Neto, A., and Ceres, R. (2012). Assistive mobility devices focusing on smart walkers: Classification and review. Robotics and Autonomous Systems, 60(4):548–562.

[Masoud, 2012] Masoud, A. A. (2012). A harmonic potential approach for simultaneous planning and control of a generic UAV platform. Journal of Intelligent and Robotic Systems, (1-4):153–173.

[Mastellone et al., 2008] Mastellone, S., Stipanovic, D., Graunke, C., Intlekofer, K., and Spong, M. W. (2008). Formation control and collision avoidance for multi-agent non-holonomic systems: Theory and experiments. The International Journal of Robotics Research, 27:107–126.

[Mastellone et al., 2007] Mastellone, S., Stipanovic, D., and Spong, M. (2007). Remote formation control and collision avoidance for multi-agent nonholonomic systems. In IEEE International Conference on Robotics and Automation, 1062–1067, Italy.

[Mataric, 1992] Mataric, M. J. (1992). Minimizing complexity in controlling a mobile robot population. In IEEE International Conference on Robotics and Automation, 830–835, Nice-France.

[Mataric et al., 1995] Mataric, M. J., Nilsson, M., and Simsarian, K. (1995). Cooperative multi-robots box-pushing. In IEEE International Conference on Intelligent Robots and Systems, volume 3, 556–561.

[Mathews et al., 2015] Mathews, N., Valentini, G., Christensen, A. L., O'Grady, R., Brutschy, A., and Dorigo, M. (2015). Spatially targeted communication in decentralized multirobot systems. Autonomous Robots, 38(4):439–457.

[Memon and Bilal, 2015] Memon, W. A. and Bilal, M. (2015). Fully autonomous flammable gases (MethaneGas) sensing and surveillance robot. In International Conference on Artificial Intelligence, Energy and Manufacturing Engineering, Dubai.

[Mesbahi and Hadaegh, 1999] Mesbahi, M. and Hadaegh, F. (1999). Formation flying control of multiple spacecraft via graphs, matrix inequalities, and switching. In Proceedings of the IEEE International Conference on Control Applications, volume 2, 1211–1216.

[Minguez et al., 2008] Minguez, J., Lamiraux, F., and Laumond, J.-P. (2008). Handbook of Robotics, chapter Motion Planning and Obstacle Avoidance, 827–852.

[Morin and Samson, 2009] Morin, P. and Samson, C. (2009). Control of nonholonomic mobile robots based on the transverse function approach. Transaction on Robotics, 25:1058–1073.

[Mouad, 2014] Mouad, M. (2014). Control and management architecture for distributed autonomous Systems: Application to multiple mobile vehicles based platform. PhD thesis, Balise Pascal University.

[Mouad et al., 2012] Mouad, M., Adouane, L., Khadraoui, D., and Martinet, P. (2012). Mobile robot navigation and obstacles avoidance based on planning and re-planning algorithm. In 10th International IFAC Symposium on Robot Control (SYROCO'12), Dubrovnik - Croatia.

[Mouad et al., 2010] Mouad, M., Adouane, L., Schmitt, P., Khadraoui, D., Gateau, B., and Martinet, P. (2010). Multi-agent based system to coordinate mobile teamworking robots. In 4th Companion Robotics Workshop, Brussels, Belgium.

[Mouad et al., 2011a] Mouad, M., Adouane, L., Schmitt, P., Khadraoui, D., and Martinet, P. (2011a). Control architecture for cooperative mobile robots using multi-agent based coordination approach. In CAR'11, 6th National Conference on "Control Architectures of Robots", Grenoble, France.

[Mouad et al., 2011b] Mouad, M., Adouane, L., Schmitt, P., Khadraoui, D., and Martinet, P. (2011b). Mas2car architecture: Multi-agent system to control and coordinate teamworking robots. In ICINCO'11, 8th International Conference on Informatics in Control, Automation and Robotics, Nederlands.

[Muñoz, 2003] Muñoz, A. (2003). Coopération située : Une approche constructiviste de la conception de colonies de robots. PhD thesis, Université Pierre et Marie Curie, Paris VI.

[Murata and Kurokawa, 2012] Murata, S. and Kurokawa, H. (2012). Self-Organizing Robots, volume 77 of Tracts in Advanced Robotics. Springer.

[Murphy, 2012] Murphy, R. R. (2012). A decade of rescue robots. In IEEE/RSJ International Conference onIntelligent Robots and Systems, 5448–5449.

[Murray, 2007] Murray, R. (2007). Recent research in cooperative control of multi-vehicle systems. J. Dyn. Sys., Meas., Control, 129(5):571–583.

[NHTSA, 2013] NHTSA (2013). National Highway Traffic Safety Administration: Preliminary Statement of Policy Concerning Automated Vehicles - USA.

[Noreils, 1993] Noreils, F. R. (1993). Toward a robot architecture integrating cooperation between mobile robots: Application to indoor environment. International Journal of Robotics Research, 12(1):79–98.

[Ogren et al., 2002] Ogren, P., Fiorelli, E., and E., L. N. (2002). Formations with a mission: Stable coordination of vehicle group maneuvers. In 15th International Symposium on Mathematical Theory of Networks and Systems.

[Ogren and Leonard, 2005] Ogren, P. and Leonard, N. E. (2005). A convergent dynamic window approach to obstacle avoidance. IEEE Transactions on Robotics, 21(2):188–195.

[Ordonez et al., 2008] Ordonez, C., Jr., E. G. C., Selekwa, M. F., and Dunlap, D. D. (2008). The virtual wall approach to limit cycle avoidance for unmanned ground vehicles. Robotics and Autonomous Systems, 56(8):645–657.

[Ota, 2006] Ota, J. (2006). Multi-agent robot systems as distributed autonomous systems. Advanced Engineering Informatics, 20:59–70.

[Papadimitratos et al., 2008] Papadimitratos, P., Poturalski, M., Schaller, P., Lafourcade, P., Basin, D., Capkun, S., and Hubaux, J.-P. (2008). Secure Neighborhood Discovery: A Fundamental Element for Mobile Ad Hoc Networking. IEEE Communications Magazine, 46(2):132–139.

[Parker, 1996] Parker, L. (1996). On the design of behavior-based multi-robot teams. Journal of Advanced Robotics, 10:547–578.

[Parker, 2009] Parker, L. (2009). Encyclopedia of Complexity and System Science [online], chapter Path Planning and Motion Coordination in Multiple Mobile Robot Teams, 5783–5799. Springer.

[Parker, 1998] Parker, L. E. (1998). ALLIANCE: An architecture for fault tolerant multi-robot cooperation. IEEE Transactions on Robotics and Automation, 14(2):220–240.

[Parker, 1999] Parker, L. E. (1999). Adaptive heterogeneous multi-robot teams. In Neurocomputing, special issue of NEURAP '98: Neural Networks and Their Applications, volume 28, 75–92.

[Pesterev, 2012] Pesterev, A. V. (2012). Stabilizing control for a wheeled robot following a curvilinear path. In 10th International IFAC Symposium on Robot Control. Croatia.

[Pirjanian, 2000] Pirjanian, P. (2000). Multiple objective behavior-based control. Journal of Robotics and Autonomous Systems, 31(1):53–60.

[Pirjanian and Mataric, 2001] Pirjanian, P. and Mataric, M. (2001). Multiple objective vs. fuzzy behavior coordination. In Driankov, D. and Saffiotti, A., editors, Fuzzy Logic Techniques for Autonomous Vehicle Navigation, volume 61 of Studies in Fuzziness and Soft Computing, 235–253. Physica-Verlag HD.

[Pivtoraiko and Kelly, 2009] Pivtoraiko, M. and Kelly, A. (2009). Fast and feasible deliberative motion planner for dynamic environments. In International Conference on Robotics and Automation. Workshop on Planning in Dynamic Environments.

[Porrill, 1990] Porrill, J. (1990). Fitting ellipses and predicting confidence envelopes using a bias corrected kalman filter. Image and Vision Computing, 8(1):37–41.

[Ranganathan and Koenig, 2003] Ranganathan, A. and Koenig, S. (2003). A reactive robot architecture with planning on demand. In IEEE/RSJ International Conference on Intelligent Robots and Systems, 1462–1468, Las Vegas, Nevada, USA.

[Ridao et al., 1999] Ridao, P., Batlle, J., Amat, J., and Roberts, G. N. (1999). Recent trends in control architectures for autonomous underwater vehicles. Int. J. Systems Science, 30(9):1033–1056.

[Rimon and Daniel.Koditschek, 1992] Rimon, E. and Daniel.Koditschek (1992). Exact robot navigation using artificial potential fields. IEEE Transactions on Robotics and Automation, 8(5):501–518.

[Rodriguez, 2014] Rodriguez, G. S. (2014). Detection based on laser range data and segmentation for an autonomous vehicle. Master's thesis, MASTER II: ERASMUS between Valladolid (Espagne) and UBP Universities.

[Rouff and Hinchey, 2011] Rouff, C. and Hinchey, M. (2011). Experience from the DARPA Urban Challenge. Springer Publishing Company, Incorporated.

[Royer et al., 2007] Royer, E., Lhillier, M., Dhome, M., and Lavest, J.-M. (2007). Monocular vision for mobile robot localization and autonomous navigation. Journal of Computer Vision, 74(3):237–260.

[SAE, 2015] SAE (2015). International J3016: Taxonomy and Definitions for Terms Related to On-Road Motor Vehicle Automated Driving Systems. http://standards.sae.org/j3016_201401, consulted January 2015.

[Saffiotti et al., 1993] Saffiotti, A., Ruspini, E., and Konolige, K. (1993). Robust execution of robot plans using fuzzy logic. In Springer-Velag, editor, Fuzzy Logic in Artificial Intelligence: IJCAI'93 Workshop, 24–37, Chambéry-France.

[Samson, 1995] Samson, C. (1995). Control of chained systems: Application to path following and time-varying point-stabilization of mobile robots. IEEE Transactions on Automatic Control, 40(1):64–77.

[Seeni et al., 2010] Seeni, A., Schäfer, B., and Hirzinger, G. (2010). Robot mobility systems for planetary surface exploration-state-of-the-art and future outlook: A literature survey. INTECH Open Access Publisher.

[Segundo and Sendra, 2005] Segundo, F. S. and Sendra, J. R. (2005). Degree formulae for offset curves. Journal of Pure and Applied Algebra, 195(3):301–335.

[Sezen, 2011] Sezen, B. (2011). Modeling automated guided vehicle systems in material handling. Otomatiklestirilmi Rehberli Arac Sistemlerinin Transport Tekniginde Modellemesi, Dou Universitesi Dergisi, 4(2):207–216.

[Shames et al., 2011] Shames, I., Deghat, M., and Anderson, B. (2011). Safe formation control with obstacle avoidance. In IFAC World Congress, Milan - Italy.

[Siciliano and Khatib, 2008] Siciliano, B. and Khatib, O., editors (2008). Springer Handbook of Robotics, Part E-34. Springer.

[Siegwart et al., 2011] Siegwart, R., Nourbakhsh, I., and Scaramuzza, D. (2011). Introduction to Autonomous Mobile Robots. Intelligent robotics and autonomous agents, MIT Press. MIT Press.

[Sigaud and Gérard, 2000] Sigaud, O. and Gérard, P. (2000). The use of roles in a multiagent adaptive simulation. In Proceedings of the 14th European Conference in Artificial Intelligence, Workshop on Balancing reactivity and Social Deliberation in Multiagent Systems, Berlin, Germany.

[Simonin, 2001] Simonin, O. (2001). Le modèle satisfaction-altruisme : Coopération et résolution de conflits entre agents situés réactifs, application à la robotique. PhD thesis, Université Montpellier II.

[Simpkins and Simpkins, 2014] Simpkins, A. and Simpkins, C. (2014). Rescue robotics [on the shelf]. Robotics Automation Magazine, IEEE, 21(4):108–109.

[Siouris, 2004] Siouris, G. M. (2004). Missile Guidance and Control Systems. Springer-Verlag.

[Soltan et al., 2011] Soltan, R. A., Ashrafiuon, H., and Muske, K. R. (2011). Ode-based obstacle avoidance and trajectory planning for unmanned surface vessels. Robotica, 29:691–703.

[Speers and Jenkin, 2013] Speers, A. and Jenkin, M. (2013). Diver-based control of a tethered unmanned underwater vehicle. In Ferrier, J.-L., Gusikhin, O. Y., Madani, K., and Sasiadek, J. Z., editors, ICINCO (2), 200–206. SciTePress.

[Springer, 2013] Springer, P. J. (2013). Military Robots and Drones: A Reference Handbook. ABC-CLIO.

[Stoeter et al., 2002] Stoeter, S. A., Rybski, P. E., Stubbs, K. N., McMillen, C. P., Gini, M., Hougen, D. F., and Papanikolopoulos, N. (2002). A robot team for surveillance tasks: Design and architecture. Robotics and Autonomous Systems, 40(2-3):173–183.

[Szczerba et al., 2000] Szczerba, R. J., Galkowski, P., Glickstein, I. S., and Ternullo, N. (2000). Robust algorithm for real-time route planning. IEEE Transactions on Aerospace and Electronics Systems, 36:869–878.

[Takahashi et al., 2010] Takahashi, M., Suzuki, T., Shitamoto, H., Moriguchi, T., and Yoshida, K. (2010). Developing a mobile robot for transport applications in the hospital domain. Robotics and Autonomous Systems, 58(7):889–899. Advances in Autonomous Robots for Service and Entertainment.

[Tang et al., 2006] Tang, H., Song, A., and Zhang, X. (2006). Hybrid behavior coordination mechanism for navigation of reconnaissance robot. In IEEE/RSJ International Conference on Intelligent Robots and Systems, 1773–1778, Beijing, China.

[Thrun, 2011] Thrun, S. (2011). IEEE Spectrum, How Google's Self-Driving Car Works, http://spectrum.ieee.org/automaton/robotics/artificial-intelligence/how-google-self-driving-car-works.

[Thrun et al., 2005] Thrun, S., Burgard, W., and Fox, D. (2005). Probabilistic Robotics (Intelligent Robotics and Autonomous Agents). The MIT Press.

[Thrun et al., 2007] Thrun, S., Montemerlo, M., Dahlkamp, H., Stavens, D., Aron, A., Diebel, J., Fong, P., Gale, J., Halpenny, M., Hoffmann, G., et al. (2007). Stanley: The robot that won the DARPA Grand Challenge. In The 2005 DARPA Grand Challenge, 1–43. Springer.

[Toibero et al., 2007] Toibero, J., Carelli, R., and Kuchen, B. (2007). Switching control of mobile robots for autonomous navigation in unknown environments. In IEEE International Conference on Robotics and Automation, 1974–1979, Roma, Italy.

[Toutouh et al., 2012] Toutouh, J., Garcia-Nieto, J., and Alba, E. (2012). Intelligent OLSR routing protocol optimization for vanets. IEEE Transactions on Vehicular Technology, 61(4):1884–1894.

[Van den Berg and Overmars, 2005] Van den Berg, J. and Overmars, M. (2005). Roadmap-based motion planning in dynamic environments. IEEE Transactions on Robotics, 21:885–897.

[VANET, 2015] VANET (2015). VANET (Vehicular ad hoc network). http:// en.wikipedia.org/wiki/Vehicular_ad_hoc_network, consulted January 2015.

[Vaz et al., 2010] Vaz, D. A., Inoue, R. S., and Grassi Jr., V. (2010). Kinodynamic motion planning of a skid-steering mobile robot using RRTs. In Proceedings of the Latin American Robotics Symposium and Intelligent Robotics Meeting, LARS '10, 73–78, Sao Bernardo do Campo, Brazil. IEEE Computer Society.

[Vilca et al., 2012a] Vilca, J., Adouane, L., Benzerrouk, A., and Mezouar, Y. (2012a). Cooperative on-line object detection using multi-robot formation. In 7th National Conference on Control Architectures of Robots (CAR'12), Nancy - France.

[Vilca et al., 2013a] Vilca, J., Adouane, L., and Mezouar, Y. (2013a). Reactive navigation of mobile robot using elliptic trajectories and effective on-line obstacle detection. Gyroscopy and Navigation, 4(1):14–25. Springer Verlag, Russia ISSN 2075 1087.

[Vilca et al., 2012b] Vilca, J.-M., Adouane, L., and Mezouar, Y. (2012b). On-line obstacle detection using data range for reactive obstacle avoidance. In 12th International Conference on Intelligent Autonomous System (IAS-12), Jeju Island - Korea. Published after in Advances in Intelligent Systems and Computing, 193, 2013, 3-13.

[Vilca et al., 2012c] Vilca, J.-M., Adouane, L., and Mezouar, Y. (2012c). Robust on-line obstacle detection using data range for reactive navigation. In 10th International IFAC Symposium on Robot Control (SYROCO'12), Dubrovnik - Croatia.

[Vilca et al., 2014] Vilca, J.-M., Adouane, L., and Mezouar, Y. (2014). Adaptive leader-follower formation in cluttered environment using dynamic target reconfiguration. In Springer Tracts in Advanced Robotics, from International Symposium on Distributed Autonomous Robotic Systems, DARS 2014, Daejeon - Korea.

[Vilca et al., 2015a] Vilca, J.-M., Adouane, L., and Mezouar, Y. (2015a). A novel safe and flexible control strategy based on target reaching for the navigation of urban vehicles. Robotics and Autonomous Systems (RAS), 70:215–226.

[Vilca et al., 2015b] Vilca, J.-M., Adouane, L., and Mezouar, Y. (2015b). Optimal multi-criteria waypoint selection for autonomous vehicle navigation in structured environment. Journal of Intelligent & Robotic Systems (JIRS), 1–24.

[Vilca et al., 2013b] Vilca, J.-M., Adouane, L., Mezouar, Y., and Lebraly, P. (2013b). An overall control strategy based on target reaching for the navigation of a urban electric vehicle. In IEEE/RSJ, IROS'13, International Conference on Intelligent Robots and Systems, Tokyo-Japan.

[Vilca Ventura, 2015] Vilca Ventura, J. M. (2015). Safe and Flexible Hybrid Control Architecture for the Navigation in Formation of a Group of Vehicles. PhD thesis, Balise Pascal University.

[Vine et al., 2015] Vine, S. L., Zolfaghari, A., and Polak, J. (2015). Autonomous cars: The tension between occupant experience and intersection capacity. Transportation Research Part C: Emerging Technologies, 52:1–14.

[Voth, 2004] Voth, D. (2004). A new generation of military robots. Intelligent Systems, IEEE, 19(4):2–3.

[Walter, 1953] Walter, W. G. (1953). The Living Brain. W. W. Norton, New York.

[Walton and Meek, 2005] Walton, D. J. and Meek, D. S. (2005). A controlled clothoid spline. Computers & Graphics, 29(3):353–363.

[Wang, 2011] Wang, Zhiying, D. X. R. A. G. A. (2011). Mobility analysis of the typical gait of a radial symmetrical six-legged robot. Mechatronics, 21(7):1133–1146.

[Wang and Liua, 2008] Wang, M. and Liua, J. N. (2008). Fuzzy logic-based real-time robot navigation in unknown environment with dead ends. Robotics and Autonomous Systems, 56(7):625–643.

[Welzl, 1991] Welzl, E. (1991). Smallest enclosing disks (balls and ellipsoids). In Results and New Trends in Computer Science, 359–370. Springer-Verlag.

[Wilber, 1972] Wilber, B. M. (1972). A shakey primer. Technical report, Stanford Research Institute, 333 Ravenswood Ave, Menlo Park, CA 94025.

[Williams, 1988] Williams, M. (1988). PROMETHEUS: The European research programme for optimising the road transport system in Europe. In IEEE Colloquium on Driver Information, 1/1–1/9.

[Wu et al., 2014] Wu, F., Liu, S., and Mu, W. J. (2014). A tracked robot for complex environment detecting. In Applied Mechanics and Materials, volume 670, 1389–1392. Trans Tech Publ.

[Yamada and Saito, 2001] Yamada, S. and Saito, J. (2001). Adaptive action selection without explicit communication for multirobot box-pushing. IEEE Transaction on Systems, Man and Cybernetics, Part C: Application And Reviews, 31(3):398–404.

[Yeomans, 2010] Yeomans, G. (2010). Autonomous Vehicles: Handing over Control: Risks and Opportunities in Insurance. Lloyd's exposure management edition.

[Yoshikawa, 2010] Yoshikawa, T. (2010). Multifingered robot hands: Control for grasping and manipulation. Annual Reviews in Control, 34(2):199–208.

[Zapata et al., 2004] Zapata, R., Cacitti, A., and Lepinay, P. (2004). DVZ-based collision avoidance control of non-holonomic mobile manipulators. JESA, European Journal of Automated Systems, 38(5):559–588.

[Ze-su et al., 2012] Ze-su, C., Jie, Z., and Jian, C. (2012). Formation control and obstacle avoidance for multiple robots subject to wheel-slip. International Journal of Advanced Robotic Systems, 9:1–15.

[Zefran and Burdick, 1998] Zefran, M. and Burdick, J. W. (1998). Design of switching controllers for systems with changing dynamics. In IEEE Conference on Decision and Control CDC'98, volume 2, 2113–2118, FL, USA.

[Zhang, 1997] Zhang, Z. (1997). Parameter estimation techniques: A tutorial with application to conic fitting. Image and Vision Computing, 15:59 – 76.

[Ziegler et al., 2008] Ziegler, J., Werling, M., and Schroeder, J. (2008). Navigating car-like robots in unstructured environment using an obstacle sensitive cost function. In Proc. IEEE Intelligent Vehicle Sympsium (IV), 787–791, Netherlands.

Index